Cambridge Studies in Social Anthropology

General Edit

DATE DUE

46

STRUCTUR

For a complete series list, see the end of this book.

Structural models in anthropology

PER HAGE
Department of Anthropology
University of Utah

FRANK HARARY
Department of Mathematics
University of Michigan

CAMBRIDGE UNIVERSITY PRESS

Cambridge
London   New York   New Rochelle
Melbourne   Sydney

CAMBRIDGE UNIVERSITY PRESS
Cambridge, New York, Melbourne, Madrid, Cape Town, Singapore, São Paulo

Cambridge University Press
The Edinburgh Building, Cambridge CB2 8RU, UK

Published in the United States of America by Cambridge University Press, New York

www.cambridge.org
Information on this title: www.cambridge.org/9780521253222

First published 1983
Re-issued in this digitally printed version 2007

A catalogue record for this publication is available from the British Library

Library of Congress Cataloguing in Publication data
Hage, Per, 1935–
Structural models in anthropology.
(Cambridge studies in social anthropology; 46)
Bibliography: p.
Includes index.
1. Structural anthropology. 2. Graph theory.
I. Harary, Frank. II. Title. III. Series:
Cambridge studies in social anthropology; no. 46.
GN362.H33 1983   306'.01'5115   83–7552

ISBN 978-0-521-25322-2 hardback
ISBN 978-0-521-27311-4 paperback

*To our parents
and their memory*

We may classify objects according to their matter; as
wooden things, iron things, silver things, ivory things,
etc. But classification according to structure is generally
more important. And it is the same with ideas.

Charles Sanders Peirce,
Letter to Signor Calderoni

Contents

Contents

Foreword

Of the making of "structures" there is no end. So might any onlooker think when surveying the intellectual fashions that have enlivened the development of the social sciences during the last 75 years. The term has been used for a confusing succession of notions and concepts that have shared very little with one another except the label "structure" itself. Anthropology has been powerfully influenced by Radcliffe-Brown, Parsons, and Lévi-Strauss, each offering his own distinctive structural road to intellectual enlightenment. Parsons has had most of his following in sociology, but so too have Blalock and Duncan, with their very different understanding of what is meant by structure. The liveliest structuralist controversies have erupted on the borders of social science, as conventionally defined, over the structural Marxism of Althusser, structural linguistics, and above all, structural analyses of literature.

Our onlooker, if he or she is tidy-minded and likes ideas and practices to be pigeonholed unambiguously, might regret that so many different people, without consulting one another, have taken on the role of Humpty Dumpty and declared the word "structure" to mean whatever each wanted it to mean. Intellectual discourse, he or she might well think, would be much more effective, and much less frustrating, if the humanities and social sciences were to use a technical vocabulary closer akin to that of the natural sciences, with no ambiguity and just the right amount of redundancy. Then there would be no doubt about what people were trying to say.

Alas, this vision of an immaculate natural science is only a naive onlooker's mirage. As for the humanities and social sciences, they are destined to remain permanently confused. For not only do they continually enrich the language of everyday speech with their own neologisms; they also draw on the fuzzy ordinary world, rather than on some neoclassical word factory, for terms that they can refine and provide with precise denotations. Attempts to confront, between the covers of a book, the diverse meanings of some much-used label serve mainly to chart the extent of confusion rather than to end it. For example, the symposium *Sens et usages du term structure dans les sciences humaines et sociales,* edited by Roger Bastide, provides a fascinating catalogue of diversity but offers no prospect of consensus. There are, I believe, important advantages in maintaining the permeability of the frontier between the languages of everyday life and social science, but there is also a price to be paid for it.

In 1965, Harary, Norman, and Cartwright published *Structural models: an introduction to the theory of directed graphs*. It is these sorts of structures that Hage and Harary discuss in the present volume, and it is these that form the content of the specialism now coming to be known as structural analysis. This kind of structuralism has grown out of the study of social networks and derives the majority of its analytical tools from graph theory, a branch of pure mathematics. Although many social scientists have contributed to the rise of structural analysis, many others have watched with somewhat skeptical interest. The specialism has been accused of failing to live up to its grandiose claims and of being excessively concerned with its esoteric techniques for their own sake rather than for their value in explaining social phenomena. Some critics have maintained that the anthropologist or sociologist who hopes to discover what goes on in the real world can gain nothing more from network analysis and graph theory than an unenlightening and unnecessarily complicated technical vocabulary in which to make imprecise quantitative statements about things we know already.

There has, in my view, been some force in this criticism, for in a rapidly developing specialty ideas do often race ahead of applications; progress would be slower if they did not do so. But now Hage and Harary have given us the answer to these critics. For here in this book they demonstrate with admirable clarity, and with an impressive range of illustrations, how the concepts and, more important, the theorems and techniques of graph theory can be applied to ordinary ethnographic evidence. They show convincingly that this application can yield results that could not have been obtained by unassisted common sense, results that add significantly to our understanding of the social and cultural processes taking place in the real world.

Hage and Harary have, in a sense, routinized or domesticated the analytical procedures of graph theory for use by practicing social scientists, anthropologists, and others. Several years ago, I drew a distinction between the use of the notion of social network as a metaphor and as an analytic tool. Unfortunately, most of the examples I could find at that time showed graph theory being used in quite a rough-and-ready fashion, as it were, for bush carpentry rather than for cabinetmaking. Here in this new book we have at last a comprehensive range of examples of graph theory being applied to data from the real world with the elegance and precision we rightly expect from pure mathematics. Yet Hage and Harary write with the innumerate and mathematically phobic social scientist clearly in their sights, so that no previous acquaintance with graph theory is needed. Here indeed is an opportunity for conquering that phobia which still hinders the work of so many social scientists.

We probably have to resign ourselves to living with confusion about what is meant by structuralism. But within this particular version of structural analysis, there is no excuse for confusion about what we mean by terms that are clearly technical. Regrettably, there is still a good deal of variation in the way in which different writers use terms that are derived from graph theory; indeed, some of the confusion stems from the pure mathematicians themselves. Hage and Harary provide us with a full and consistent technical vocabulary, and show us how to apply it in practical analysis. Let us hope that their usage will become generally accepted among social scientists. When notions taken from structural analysis pass into common speech (maybe with the next 20 years), we can expect confusion to grow again. But let's enjoy a Cartesian breathing space while we can.

J. A. Barnes

University of New England
Armidale, Australia
24 February 1983

Acknowledgments

The idea for this book was conceived in 1980–1, when the first author was a Visiting Fellow at Robinson College and the second author an Overseas Fellow at Churchill College, Cambridge. At that time, both of us enjoyed the kind hospitality of Professor John A. Barnes, chairman of the university's Social and Political Science Committee and long an eloquent advocate of numeracy in the social sciences. We are deeply grateful to Professor Barnes and to the linguist Dr. Susan McKay, for their critical and indispensable comments on the first draft. We are also most appreciative of the helpful remarks made by readers of individual chapters. We feel fortunate to have encountered, during our recent peregrinations, so many colleagues in anthropology and cognate disciplines who have encouraged and contributed to our project. Finally, we wish to thank Ursula Hanly for far exceeding the necessary by producing an impeccable typescript and precise and beautifully drawn figures. The first author acknowledges the essential and timely support of the University of Utah Research Committee for this collaborative venture in structural analysis. During the completion of the final version of the manuscript in the fall of 1982, the second author held the Ulam Chair of Mathematics at the University of Colorado.

We wish to thank the following for permission to reproduce the figures listed below:

The Royal Anthropological Institute for Fig. 1.8 and for the drawing on the cover from A. B. Deacon, "Geometrical drawings from Malekula and other islands of the New Hebrides," *Journal of the Royal Anthropological Institute,* vol. 64 (1934); Clarendon Press, Oxford, for Fig. 1.9(a) from N. L. Biggs, E. K. Lloyd, and R. J. Wilson, *Graph theory 1736–1936* (1976); S. H. Riesenberg and the Polynesian Society for Fig. 2.1 from "The organisation of navigational knowledge on Puluwat," *Journal of the Polynesian Society,* vol. 81 (1972); N. D. C. Hammond and Methuen for Fig. 2.18 from "Locational models and the site of Lubaatún: A classic Maya centre" in D. L. Clarke (ed.), *Models in archaeology* (1972); Prentice-Hall for Figure 3.9 from C. Flament, *Applications of graph theory to group structure* (1963); K. E. Read and the *Journal of Anthropological Research* for Fig. 3.14 from "Cultures of the Central Highlands, New Guinea,"

Acknowledgments

Southwestern Journal of Anthropology, vol. 10 (1954); C. R. Hallpike and the Royal Anthropological Institute for Figs. 3.17 and 3.18 from "The principles of alliance formation between Konso towns," *Man* (n.s.) vol. 5 (1970); Holt, Rinehart and Winston for Fig. 4.20 from W. A. Lessa, *Ulithi: A Micronesian design for living* (1966); J. Needham for Fig. 4.21 from *Science and civilisation in China,* vol. 2 (1956), Cambridge University Press; University of Minnesota Press for Fig. 5.1 from M. Levison, R. G. Ward, and J. W. Webb, *The settlement of Polynesia* (1973); M. Marriott and the Wenner-Gren Foundation for Anthropological Research for Figs. 5.7 and 5.8 from "Caste ranking and food transactions: A matrix analysis" in M. Singer and B. S. Cohn (eds.), *Structure and change in Indian society* (1968), Viking Fund Publications in Anthropology, no. 47; C. Lévi-Strauss and Harper & Row for Fig. 6.1 from *From honey to ashes* (1973); P. Doreian and Gordon and Breach for Figs. 7.1 and 7.5 from "On the connectivity of social networks," *Journal of Mathematical Sociology,* vol. 3 (1974); the *Journal of Anthropological Research* for Fig. 7.8 from W. W. Zachary, "An information flow model for conflict and fission in small groups," vol. 33 (1977); D. H. Thomas and Methuen for Figs. 7.12 and 7.13 from "A computer simulation model of Great Basin Shoshonean subsistence and settlement patterns," in D. L. Clarke (ed.), *Models in archaeology* (1972); and W. H. Freeman for Fig. 8.5 from M. Gardner, "The combinatorial basis of the 'I Ching,' the Chinese book of divination and wisdom," *Scientific American,* vol. 230 (1), 1974. Thanks are due to L. Hahn for the cover photograph.

P.H.
F.H.

1

Graph theory and anthropology

This new mathematics (which incidentally simply gives backing to, and expands on, earlier speculative thought) teaches us that the domain of necessity is not necessarily the same as that of quantity.

Claude Lévi-Strauss, "The mathematics of man"

Anthropology is fundamentally the study of sets of social and cultural relations whose diversity and pervasiveness is illustrated by such terms as "exchange," "hierarchy," "classification," "order," "opposition," "mediation," "inversion," and "transformation." The analysis of these relations always presupposes models of some kind, implicit if not explicit, informal if not formal. The models are usually defined in ordinary language, but with results that are not always satisfactory in matters of descriptive adequacy, insight, and communicability. The question thus arises as to whether, in many contexts, mathematical formulations might not be helpful; and if so, what kind of mathematics.

Some time ago, in his essay "The mathematics of man," Lévi-Strauss emphasized the suitability of the various forms of modern mathematics as a source of structural models in anthropology:

In the past, the great difficulty has arisen from the qualitative nature of our studies. If they were to be treated quantitatively, it was either necessary to do a certain amount of juggling with them or to simplify to an excessive degree. Today, however, there are many branches of mathematics – set theory, group theory, topology, etc. – which are concerned with establishing exact relationships between classes of individuals distinguished from one another by discontinuous values, and this very discontinuity is one of the essential characteristics of qualitative sets in relation to one another and was the feature in which their alleged "incommensurability," "inexpressibility," etc., consisted. (Lévi-Strauss 1955:586)

In contrasting the old and the potentially new forms of mathematical thinking in anthropology, Lévi-Strauss observed by way of example that we should be "less concerned with the theoretical consequences of a 10 per cent increase in the population in a country having 50 million inhabitants than with the changes in structure occurring when a 'two-person household' becomes a 'three-person household'" (Lévi-Strauss 1955:586). One immediately thinks of a triangle whose points represent the members of the household and whose lines represent their relations to one another. This characterization of structural anthropology serves not only to distinguish mathematical models from statistical ones, but also evokes graph theory in particular. Since the example refers to social structure and suggests a pictorial representation, it is intuitively meaningful to most anthropologists, for it is comparable to familiar anthropological graphs such as genealogical trees or drawings of trade networks. By the term "structural models" we shall mean not only just such graphs, but go beyond the mere representation of social relations to consider as graphs a variety of empirical phenomena whose underlying structural properties can be elucidated through the application of a well-developed body of relational concepts and theorems.

Our aim is to introduce graph theory as a comprehensive structural model, or family of models, in cultural and social anthropology. Graph theory is a branch of finite mathematics that is both topological and combinatorial in nature. Because it is essentially the study of relations, graph theory is eminently suited to the description and analysis of a wide range of structures that constitute a significant part of the subject matter of anthropology, as well as of the social sciences generally. We have in mind not only social networks, whose underlying graph theoretic basis is easily recognized, but a variety of social, symbolic, and cognitive structures as well. Belief and classification systems turn out to be no less graphical than communication and exchange networks. By showing this to be so, we hope not only to provide a standard language for the use of graphical models, but also to enlarge the field of structural analysis in anthropology.

In addition to the intrinsic advantages of graph theory as a structural model, there are significant interactions between graphs and relations, matrices, duality laws, and groups that increase enormously the diversity of potential empirical applications. Thus, for example, studies of the logical properties of social structure, matrices of exchange relations, transformations in myths, and permutations in symbolic systems are commonly but often unknowingly based on graph theory. When this theoretical basis is made fully explicit, analysis is usually clarified and often enhanced. Fortunately, graph theory is also relatively self-contained. Thus this book presupposes no background in it and is readily accessible to the nonmathematical reader.

For convenience of exposition, the organization of the book is dictated more by graph theoretic than by anthropological considerations. The presentation is cumulative: Chapter 2 starts with the simplest type of graph which will be modified, extended, and reinterpreted in successive chapters to yield an array of structural models. The examples brought to bear are in each case anthropologically real, as well as diverse and suggestive of further research. They are designed to convey not only a general understanding of the models, but also a concrete idea of how to apply them. This combination of theoretical exposition and meaningful exemplification will provide a basic introduction to a large and at present explosive field of applicable mathematics.[1] We will begin that introduction with a brief informal presentation of some of the key concepts of graph theory.

Graphs informally

A graph is a structure consisting of points joined by lines. Fig. 1.1 shows an ordinary graph G which depicts a subsystem of a larger exchange system in a Papuan village (Schwimmer 1973). The points represent households, and the lines represent regular and symmetrical gifts of cooked taro given by women on behalf of each household.

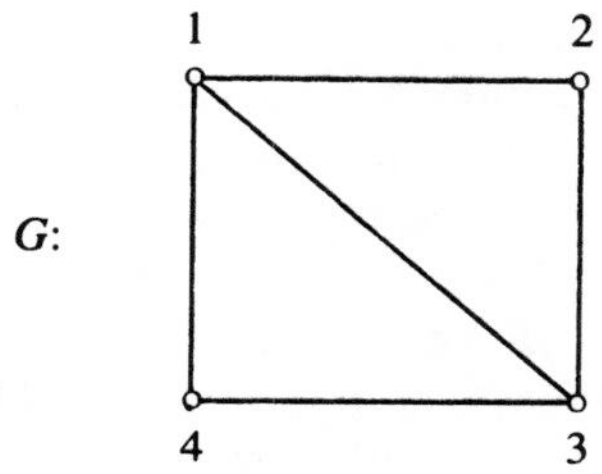

These dyadic gift relations are a sign of social intimacy, and their concatenation serves to connect indirectly all 22 households in the village, not just these 4. The indirect connections are especially significant because they are used to make requests for economic and political assistance. Thus, in Fig. 1.1 the indirect connection is a 2-step path when household 2 asks 3 (or 1) to ask 4 for a favor of some sort.

By representing a system of this type as a graph, one can study certain formal properties of social structure, together with their empirical implications. In this particular system, for example, the relative power or influence of individuals (households) depends upon the structure of mediated communication, a property that is precisely defined by the graph theoretic concept of betweenness. This refers to the frequency with which one point occurs on the paths joining all other pairs of points. Betweenness is one of the numerous definitions of centrality in a graph, and centrality, in turn, is one of a still larger set of concepts applicable to notions of communication and subordination in social systems.

We note that it makes no difference how this graph is drawn, since our only concern is with the pattern of relations, with how the points are joined by the lines. The graph of Fig. 1.2 is thus isomorphic to that of Fig. 1.1.

Fig. 1.1. A graph G of a New Guinea exchange system.

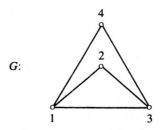

Fig. 1.2. An equivalent representation of the graph in Fig. 1.1.

[1] For more graph theory, see the text by the same title (Harary 1969). For a review of anthropological applications see Hage (1979a).

Structural models in anthropology

Social relations are not always all or none, present or absent, but may be positive or negative. By assigning + and − signs to the lines of a graph *G,* represented for convenience by solid and broken lines, respectively, we get a signed graph *S.* The one in Fig. 1.3 depicts a Melanesian system of competitive exchange based on Michael Young's (1971) monograph, *Fighting with food.* The points of S represent patrilineal clans or clan segments; the positive lines represent the relation of *fofofo,* or food friend, and the negative lines represent the relation of *nibai,* or food enemy.

In this system *nibai* make competitive gifts of forbidden food, *niune,* to each other. This food cannot be consumed; it must be passed on to a *fofofo.* Thus clan 1 gives food to clan 3, which passes it on to

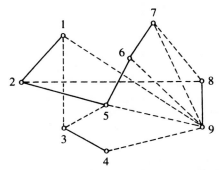

Fig. 1.3. A signed graph *S* of a Massim competitive exchange system.

clan 4; clan 9 gives to 4, which passes it on to 3. The overall pattern of food friend/enemy relations determines alignments in disputes.

There is an expectation based on theories of structural balance and clustering that relations in such systems will combine in consistent ways, or equivalently, that the group will be divisible into cliques or opposing coalitions. By modeling alliance and sentiment structures as signed graphs, we can discern consistency and contradiction in sets of relations, and in some cases we can determine, by means of simulation, how such patterns are generated.

Not all social relations are symmetric. A directed graph *D* is one in which the lines have arrows, thus permitting the representation of both asymmetric and nonsymmetric as well as symmetric relations. A theoretically interesting illustration, which shows what can happen when a projected six-person household becomes an eight-person household, comes from French literature.

Michael Oppitz (1975) in his book on structural anthropology (felicitously entitled *Notwendige Beziehungen*) introduces Lévi-Strauss's (1969) theory of marriage exchange by showing that it was anticipated over 150 years earlier by the Marquis de Sade. In *Les 120 journées de sodome,* four libertines plan to celebrate an elaborate and ongoing series of orgies at a castle owned by one of them. In order to solidify their relations, one member of the group, the Duke of Blangis, proposes a triple marriage alliance in which he will give his daughter Julie to his friend the Président de Curval,

4

the Président will give his daughter Adelaide to a third individual, a M. Durcet, and M. Durcet will give his daughter Constance to the Duke. This alliance having been agreed upon, the fourth member of the group, the Bishop of ———— is accommodated by becoming the spouse of all three women by giving his niece Aline (actually his daughter by his brother's, the Duke's, wife) to all three men. Thus, as Oppitz observes, Sade becomes the "first anthropological theoretician who expressly represents marriage as a system of communication."

The triple alliance among the Duke, the Président, and M. Durcet is an asymmetric form of marriage exchange that corresponds to Lévi-Strauss's generalized exchange: In the simplest case A gives a woman to B, B gives one to C, and C gives one to A. The alliance among these three individuals and the Bishop is a symmetric form of marriage exchange and corresponds to Lévi-Strauss's restricted exchange: In the simplest case A gives a woman to B, and B gives one to A.[2] The directed graph of this system, which incorporates both types of relations, symmetric and asymmetric, is shown in Fig. 1.4.

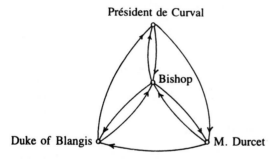

Fig. 1.4. A directed graph *D* of marriage exchange in Sade's *Les 120 journées de sodome*.

One of the most significant advantages of graph theory is that it provides diagrammatic models for the representation of diverse types of structures. For example, the logical analysis of social structure recommended some time ago by Lévi-Strauss (1963a) and subsequently attempted in areas as diverse as Basque studies (Ott 1981) and Melanesian anthropology (Kelly 1974) would be enormously facilitated by using graphs such as the one in Fig. 1.4 to model concepts like symmetry, transitivity, reflexivity.

[2]The terms "generalized exchange" and "restricted exchange" are actually complex and based on multiple criteria, as Barnes (1971) shows in his excellent discussion. For present purposes: "Generalized exchange establishes a system of operations conducted 'on credit.' A surrenders a daughter or sister to B, who surrenders one to C who, in turn will surrender one to A. This is its simplest formula" (Lévi-Strauss 1969:265). "The term 'restricted exchange' includes any system which effectively or functionally divides the group into a certain number of pairs of exchange units, so that for any one pair X-Y there is a reciprocal exchange relationship" (Lévi-Strauss 1969:146).

Actually, a graph need not be represented pictorially, although it is often convenient and intuitively helpful to do so. Instead, a matrix can be used in which each point has a row and a column and in which the entries in the cells are either 1 or 0 to show the presence or absence of a line joining a pair of points. Thus the adjacency matrix A of the graph G in Fig. 1.1 is as follows:

$$A(G) \;=\; \begin{array}{c} \\ 1 \\ 2 \\ 3 \\ 4 \end{array} \begin{array}{cccc} 1 & 2 & 3 & 4 \\ \left[\begin{array}{cccc} 0 & 1 & 1 & 1 \\ 1 & 0 & 1 & 0 \\ 1 & 1 & 0 & 1 \\ 1 & 0 & 1 & 0 \end{array}\right] \end{array}$$

A matrix conveys exactly the same information as a drawing, but has the advantage that it can be algebraically manipulated to reveal higher-order structural properties – for example, the distance between each pair of points. Thus properties such as centrality, clustering, and transitivity in graphs, signed graphs, and directed graphs can be studied just by using matrix methods. Indeed they must be studied in this way when structures become large.

If we use the term graph in its generic sense to include ordinary graphs G, signed graphs S, and directed graphs D, perhaps its most immediate interpretation is as an exchange system like one of those noted so far. However, any of these types of graphs could depict cognitive schemata where the points represent categories and the lines represent relations, say, of contrast, implication, sequencing, and so on. Or they could depict symbolic systems where the points represent sets of cultural beliefs or practices and the lines relations of transformation or symmetry.

One way to think of a transformation is to use the theory of structural duality in graphs. For every type of graph, there is a dual operation that changes it to give another graph with the same set of points and that, when applied twice, results in the original graph. This is structural duality. In the case of a directed graph D, for example, the operation, called taking the converse, consists of reversing the direction of all the arrows to get D'. A very simple example from mythology may be given. In *Naked man*, Lévi-Strauss (1981) describes inverse relations between myths concerning the loon as found on the Pacific and Atlantic coasts. One of these inversions concerns the hierarchical relationship between the loon and another bird, the diver, which Lévi-Strauss represents symbolically as in Fig. 1.5(a), and which we depict as a directed graph D and its converse D' in Fig. 1.5(b).

Besides the operation of taking the converse, there are comparable dual operations for signed graphs and for ordinary graphs called negation and complementation, respectively. The theory of structural duality provides a source of transformation rules for the analysis of symbolic systems and at the same time yields a way of disentangling various meanings of the word "opposite," which carries such a heavy semantic load in structural analysis.

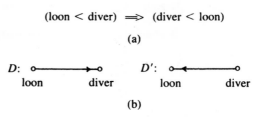

(loon < diver) $\Longrightarrow$ (diver < loon)

(a)

D: loon — diver D': loon — diver

(b)

Fig. 1.5. The converse D' of a directed graph D in mythology.

For certain types of problems in anthropology, one is concerned not only with the presence or absence of a relationship, but also with its strength or multiplexity, as is emphasized, for example, in social network studies (Mitchell 1969). Then, when empirically justified, numbers can be assigned to the lines of a graph to give a network N. In the system depicted in Fig. 1.1, for example, the ethnographer determined not only the existence of food exchanges between households, but tabulated the amount of the exchanges as well by making daily observations over a period of several months. This enabled him to discriminate first, second, and third preferential partners on the basis of the relative amounts household A gave to households B, C, and D. Fig. 1.6 shows the resulting network of the graph in Fig. 1.1; the numbering on the lines indicates the degree of preference of one exchange partner for another. By assigning numbers to the lines of a graph or directed graph, it becomes possible not only to take account of the strength of a social relation, but also to study certain aspects of group dynamics using models of information flow in capacitated networks. When the numbers are probabilities, one can study the structure of transitions be-

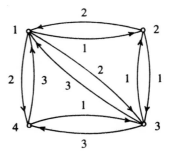

Fig. 1.6. A network N from the graph G in Fig. 1.1.

tween states of social and ecological systems using the model of a markov chain.

Graphs can represent algebraic groups as well as social groups. Although group theory has most often been applied to marriage systems, it is no less fundamental in the analysis of other forms of symbolism based on permutation relations. An example is provided by Dell Hymes (1971), who uses an informal group model to elucidate a basic genre, or semantic field, in Clackamas Chinook mythology. The genre is defined by the permutations resulting from (1) the upholding versus the violation of a social norm and (2) an adequate versus an inadequate response to an empirical situation.

These two relations yield four types, each of which has its own character-istic outcome.

Thus, for example, in the myth "Kušaydi," the murderous hero eats something about which he has been warned and dies as a result. In the myth "Seal took them to the ocean," the hero mistreats his elder brother but manages to survive many physical contests underwater by heeding Seal's advice. The outcome, however, is mixed; no accession of power or wealth, despite all the underwater accomplishments. In the myth "Seal woman and her younger brother dwelt there," the heroine insists on maintaining social propriety at the expense of ignoring a danger, with tragic consequences. In "Black Bear and Grizzly Woman and their sons," the heroes behave respectably and succeed in avenging their mother and outwitting Grizzly Woman.

Hymes characterizes his procedure by quoting Lévi-Strauss's celebrated definition of structuralist method enunciated in *Totemism* (1963b:16):

1. define the phenomenon under study as a relation between two or more terms, real or supposed;
2. construct a table of possible permutations between these terms;
3. take this table as the general object of analysis....

Hymes's model can be clearly represented by a cube graph like that in Fig. 1.7, where U and V signify the upholding versus the violation of a social norm, and A and I signify an adequate versus an inadequate response to an empirical situation. In such a graph, each permutation corresponds to a distinctive combination of the numbers 0 and 1 at the corners. This is a model of the famous Klein group evoked in the structuralist analysis of myth (Lévi-Strauss 1968, 1981). It is also a specific instance of the general model of a boolean group (Hage and Harary 1983).

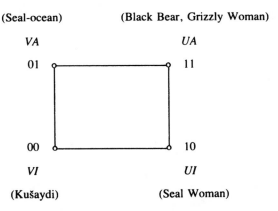

Fig. 1.7. A permutation group of Chinook myths.

Graph theory and anthropology

Group models are basic to the analysis of symmetry, a property that is not limited to physical structures, but is also a characteristic of certain forms of primitive classification. Indeed, from Piaget's point of view, the axioms that define a group are fundamental operations in the ontogenesis of intelligence. The proper evaluation of his theories and of his phylogenetic extrapolations to primitive thought, which have recently become a topic of controversy in anthropology, depends on knowing what these axioms are.

Advantages of graphical models

Graph theory offers four advantages as a structural model in anthropology. First of all, graphical models are in some sense iconic; they look like what they represent and thus, unlike algebraic models, they are "linked with reality" (Berge 1962). It is easy to understand a social or cognitive structure as a graph open to inspection and amenable to manipulation for the elucidation of its structural properties.

Second, graph theory provides exact and rich definitions of such concepts as connectivity, transformation, duality, and centrality. This language serves both to clarify common anthropological metaphors and to extend structural analysis. Thus one can sort out what is meant or what should be meant or even what could be meant by terms like "connectivity" in social networks or "permutations" in symbolic systems. At the same time, one can conceptualize completely new topics, such as "simplicity" and "complexity," in either type of structure.

Third, graph theory contains techniques for the calculation of quantitative aspects of structure – for example, reachability in a communication network, distance in a trade network, or transitivity in a status system. This is facilitated by the representation of a graph as a matrix and hence the application of simple and easily programmed algebraic manipulations. Matrices also have ethnographic advantages. As Marriott (1968) discovered in his analysis of caste relations in India, they have a certain naturalness and inevitability in the representation of complex structures.

Finally, graph theory contains theorems. By stating as a hypothesis those properties of graphs that necessarily satisfy given empirical conditions, a theorem allows us to draw conclusions about certain properties of a structure from knowledge about other properties. A simple but historically interesting example taken from Ascher and Ascher's (1981) discussion of implicit or folk mathematics will serve to illustrate.

As the Aschers observe, "mathematics arises out of and is directly concerned with the domain of thought involving the concepts of number, spatial configuration and logic." These are concerns not only of professional mathematicians, but of numerous other professional groups

9

and in fact of individuals in all cultures. One of their examples shows how a problem studied by Melanesian artists is the same one that dates the origin of graph theory in Western mathematics. They cite an early study of New Hebrides geometrical art in which A. Bernard Deacon (1934) describes a number of figures, some of which have religious, mythological, or ritual significance, and all of which have distinctive names. The figures are drawn in the earth around a framework of lines or dots. "In theory the whole should be done in a single continuous line which ends where it began; the finger should never be lifted from the ground, nor should any part of the line be traversed twice" (Deacon 1934:133). One of the simpler such drawings is "The wild cabbage," shown in Fig. 1.8.[3]

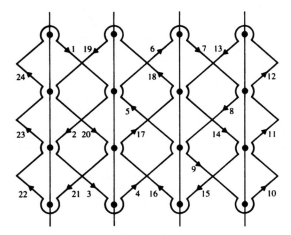

Fig. 1.8. A New Hebrides geometrical drawing (from Deacon 1934).

It is interesting to compare this drawing to the one in Fig. 1.9(a), which depicts the famous seven bridges of Königsberg and dates the origin of graph theory with the Swiss mathematician Leonhard Euler in 1736.[4] Euler (1707–1782) began the formal study of graph theory (as well as topology, which was earlier called *analysis situs*) when he settled in the negative a famous unsolved puzzle of his day called the "Königsberg Bridge Problem." As Euler himself said:

The problem, which I am told is widely known, is as follows: in Königsberg in Prussia, there is an island A, called *the Kneiphof*; the river which surrounds it is divided into two branches, as can be seen in Fig. [1.9(a)], and these branches are

[3]The cover design, which features one of these drawings, owes much to the Russian painter Andrea Morguloff.
[4]We note that Fig. 1.9(a) is taken from Euler's original paper, as reproduced in the book on the history of graph theory by Biggs, Lloyd, and Wilson (1976), and unlike many depictions, shows the bridges as they really were.

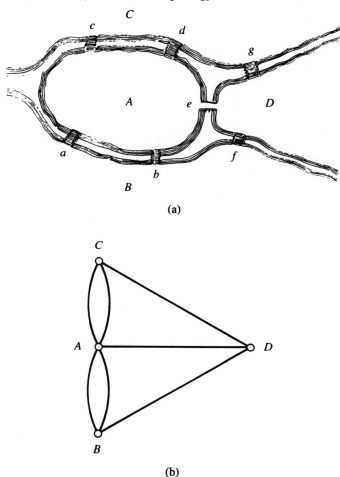

Fig. 1.9. The bridges of Königsberg (from Biggs, Lloyd, and Wilson 1976) and a graph of the Königsberg Bridge Problem.

crossed by seven bridges, *a*, *b*, *c*, *d*, *e*, *f*, and *g*. Concerning these bridges, it was asked whether anyone could arrange a route in such a way that he would cross each bridge once and only once. I was told that some people asserted that this was impossible, while others were in doubt; but nobody would actually assert that it could be done. From this, I have formulated the general problem: whatever be the arrangement and division of the river into branches, and however many bridges there be, can one find out whether or not it is possible to cross each bridge exactly once? (In Biggs, Lloyd, and Wilson 1976:3)

One can easily try to solve this problem empirically, but all attempts must be unsuccessful, for the innovative theorem of Euler proved that the task is impossible by specifying just when such a walk can be taken.

In proving that the problem is unsolvable, Euler replaced each land area by a point and each bridge by a line joining the corresponding points, thereby producing the first published graph theoretic model. This graph, actually a multigraph, as it has two lines joining some of its point pairs, is shown in Fig. 1.9(b), where the points are labeled to correspond to the four land areas of Fig. 1.9(a). Showing that the problem is unsolvable is equivalent to showing that the multigraph in Fig. 1.9(b) cannot be traversed by drawing it without removing pencil from paper.

Euler, rather than treating this specific conundrum, generalized the problem and developed a criterion for a given graph or multigraph to be so traversable – namely, that it is connected (every pair of points can reach each other), and either every point has even degree (the number of lines that emanate from a point is even) or there are exactly two points of odd degree. Although Fig. 1.9(b) is connected, not every point has even degree; in fact, all four points have odd degree. This criterion is a theorem, the first one in graph theory.

The operation of this same theorem can be seen in the New Hebrides geometric figures, which can be drawn in the way specified by Deacon precisely because each point has even degree. One can therefore accurately and succinctly characterize the basic underlying structural property of this art form by saying that whatever the particular configurations may look like or represent, they are eulerian; that is, they are connected and all points have even degree. As Ascher and Ascher emphasize, this is not an isolated case of folk or implicit mathematics. In fact, it seems to us that examples such as this one open up a new and significant field of research in anthropology.

In modern times, the Königsberg Bridge Problem has assumed a new dimension. The city of Königsberg was formerly in Prussia, but is now in the USSR and bears the name Kaliningrad. The river continues to be called the Pregel. All the bridges were destroyed during World War II, and eight bridges were later built – the original seven and one more. Consequently, the multigraph of Fig. 1.9(b) now has one additional line and hence has just two points of odd degree. Such a graph can be drawn without lifting a pencil from paper by starting at one of these points and ending at the other (although it cannot be so drawn by starting and ending at the same point).

Implicit and explicit structural models

The anthropological materials adduced to exemplify the theoretical constructs are chosen to show not just that graph theory *can* be used in anthropology, but rather that it is, to a significant degree, already *in* anthropology, and not only representationally in the form of disguised graphs, signed graphs, digraphs, and networks. Graph theory is implicit in numerous analytical concepts – for example, tree, cycle, symmetry, binary relation, isomorphism, complement, cluster, lattice, inverse, permutation group. There have also been independent discoveries of matrix methods and duality principles. In order to underline this presence, Chapters 2 through 8 all begin with an implicit anthropological use of a particular aspect of graph theory. Indeed, graph theory is such a congenial mode of anthropological thought that if Euler had not discovered it, some anthropologist might very well have.

Mathematical notions abound in anthropology. In many contexts it is sufficiently illuminating to proceed strictly at the level of metaphor and analogy. But in other contexts the mathematics really must be made explicit. If one is actually going to use matrices to analyze exchange, logical relations to define social classifications, networks to model ecological processes, or the concepts of group theory to interpret assertions concerning the operational basis of primitive thought, then for the sake of clarity, precision, and analytical power, it becomes essential to use a formal language.

Our formal presentation of structural models parallels the informal one given in this chapter. We begin by introducing ordinary graphs and providing a set of basic definitions, which are then built upon as the discussion proceeds. After obtaining signed and directed graphs and placing graphs and directed graphs in a general taxonomy of nets and relations, we turn to matrix presentations and matrix operations. Having defined the different types of graphs, we show how they may be transformed by performing dual operations on them. We then modify graphs to derive networks. Finally, we illustrate one of the interactions between graph theory and group theory by modeling boolean groups as n-cube graphs. In the Appendix, we define the concept of a mathematical model and present the axiomatic basis of the models used in this book.

2
Graphs

Diagrammatic reasoning is the only really fertile reasoning.

Charles Sanders Peirce, *The simplest mathematics*

Some of the most comprehensive, highly organized, and explicitly imparted systems of primitive classification are contained in Micronesian navigational knowledge, which includes not only oceanography and astronomy, but mythology, ritual, and magic. Navigators, called *pelu,* receive formal instruction over a number of years, on some islands – for example, Puluwat – as students of two rival navigational schools that periodically engage in exuberant intellectual contests (Gladwin 1970). So extensive is this knowledge of matters both empirical and nonempirical that chiefs undertake training as *pelu* to assure their intellectual preeminence in all domains. One device that facilitates learning is the use of mnemonics. Saul Riesenberg (1972) describes 11 types of schemata from Puluwat that aid in the memorization of topographical relations. These structures consist of points that represent places in the ocean such as islands and reefs, and lines that represent star courses between them. The learning procedure consists of enumerating sequences of locations and courses. Fig. 2.1 reproduces Riesenberg's drawing of one such mnemonic, *herekilimahacha,* the sail of Limahacha, the course of a fish or, in some places, of a legendary navigator.

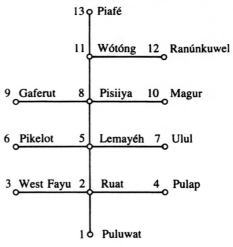

Fig. 2.1. Riesenberg's (1972) diagram of a Puluwatese mnemonic structure.

14

Graphs

Point 1, Puluwat, is the starting point; points 2, 5, 8, 11 are reefs; point 12 is a localized whale with two tails; and the remaining points are islands. One proceeds from point 1 to 2 under star x, from 2 to 3 under star y, from 3 to 2 under star z, and so on. Because one learns a star course together with its reciprocal, the traversal could be made from either end. Drawings of three additional mnemonics adapted from Riesenberg are shown in Fig. 2.2.

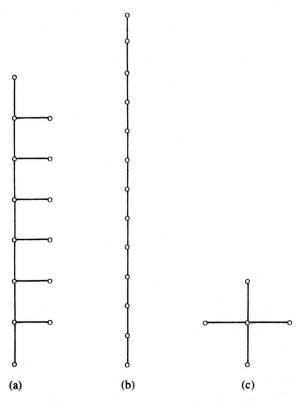

(a)   (b)   (c)

Fig. 2.2. Graphs of Puluwatese mnemonic structures.

All these structures are named and based on a visual image. All are also graphs; in particular, they are trees, the simplest type of connected graph. In fact, they are all a particular type of tree known as a caterpillar, a fact of considerable cognitive significance.

Ordinary graphs (or more simply, graphs) provide the most elementary models of cognitive and social structures. Our discussion begins with a set of definitions which are basic to all that follows. We then take up trees and analyze, from a cognitive point of view, the type of tree shown in Fig. 2.1.

15

Structural models in anthropology

The analysis accounts for the efficacy of these structures and also generalizes results that have been obtained in psychological experiments on the relative ease of learning different types of structures.

Next, we introduce a diametrically opposed type of graph known as a block, exemplified by means of a Melanesian big man exchange system. The analysis of this system illustrates the use of theorems by showing how the formally equivalent structural properties of a graph may be empirically interpreted. It also illustrates how graph theoretic concepts can be used to delineate social groups which, although not conventionally labeled, are nonetheless empirically significant. Finally, an explication of the polysemous concept of centrality in a graph has both ethnographic and archeological consequences. In the first case, it provides a model for studying the relation between mediation and power in exchange and communication systems and thus a new way of thinking about such well-known types as big man polity. In the second case, it provides a means for deciphering accessibility patterns in architectural and ceremonial layouts.

Basic definitions

Intuitively, a graph consists of a set of points, some pairs of which are joined by lines. Thus it is a kind of abstract geometry. There are precise definitions of a graph taken from logic (a symmetric irreflexive relation) and from topology (a 1-dimensional simplicial complex), but we will not build on these approaches here. A graph can also be algebraically characterized as a square symmetric matrix of zeros and ones with only zeros on the main diagonal. Our formal definition is set theoretic.

A *graph G* consists of a finite nonempty set $V = V(G)$ of *p points* together with a prescribed set E of q unordered pairs of distinct points of V. Each pair $e = \{u, v\}$ of points in E is a *line*[1] of G, and e is said to *join u* and *v*. We also write $e = uv$ and say that u and v are *adjacent points*; point u and line e are *incident* with each other, as are v and e. If two distinct lines are incident with a common point, then they are *adjacent lines*. A graph with p points and q lines is called a (p, q) *graph*. The $(1, 0)$ graph is called *trivial*, mainly in order to rule it out by specifying that a graph is to be *nontrivial*.

It is customary to represent a graph by means of a diagram and to refer to it as the graph. Thus in the graph G of Fig. 2.3, the points u and v are adjacent, but u and w are not; lines a and b are adjacent, but a and c are not. Although the lines a and c intersect in the diagram, their intersection is not a point of the graph. As we will see later, there are equivalent ways in which to represent a graph, depending on one's purpose.

[1]The following list of synonyms is sometimes used in the literature, not always with the indicated pairs:

| point | vertex | node | junction | 0-simplex | element |
| line | edge | arc | branch | 1-simplex | element |

16

Graphs

A *labeling* of a graph G with p points is an assignment of numbers 1 to p to its points. The *degree* of a point v_i in a graph G, denoted *deg* v_i or d_i, is the number of lines incident with v_i. In Fig. 2.4 there are two labeled graphs, G_1 and G_2, each of which has four points. Here G_1 is the complete graph, with four points, and G_2 is obtained from it by removing any one line. All points of G_1 have the same degree, 3; such a graph is *regular*. On the other hand, G_2 has two points of degree 2 and two of degree 3.

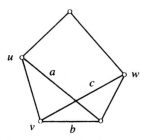

Fig. 2.3. A graph to illustrate adjacency.

G_1:
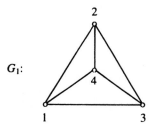

G_2:
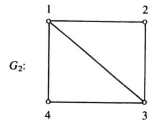

Fig. 2.4. Two labeled graphs.

In Chapter 1, Euler's solution of the Königsberg Bridge Problem was given. The basic though elementary observation he required to do this is that for any (p, q) graph G whose points have degrees, $d_1, d_2, \ldots, d_p$,

$$\sum d_i = 2q \ .$$

Thus, the sum of the degrees of the points of a graph is twice the number of lines. The simple proof of this assertion is that each line contributes 2 to this sum: 1 for each of its points. It is an immediate corollary that the number of points of odd degree is even. This has been expressed in terms of human relations by saying that the number of people who have shaken hands (or performed any other symmetric act) with an odd number of people is even.

A *walk* of a graph G is an alternating sequence of points and lines $v_0, e_1, v_1, \ldots, v_{n-1}, e_n, v_n$, beginning and ending with points, in which each line is incident with the two points immediately preceding and following it. This walk *connects* v_0 and v_n, and may also be denoted $v_0 v_1 v_2 \ldots v_n$ (or more briefly simply by 0 1 2 ... n when there is no confusion); it is sometimes called a v_0-v_n walk. It is *closed* if $v_0 = v_n$, and is *open* otherwise. It is a *trail* if all the lines are distinct (different) and a *path* if all the points (and thus necessarily all the lines) are distinct. If $n > 3$ and the walk has distinct lines and also distinct points except for its end points $v_0 = v_n$, it is a *cycle*.

17

In the labeled graph G of Fig. 2.5, 1 2 5 2 3 is a walk which is not a trail, and 1 2 5 4 2 3 is a trail which is not a path; 1 2 5 4 is a path; and 2 4 5 3 2 is a cycle.

The *length* of a path is the number of lines in it. The *distance* between two points v_i and v_j, denoted $d(v_i, v_j)$ or d_{ij}, is the length of any shortest path or *geodesic* that joins them. Unfortunately, a limited number of letters must serve many needs; thus d_i is the degree of the ith point, whereas d_{ij} is the distance from it to the jth point. In the graph G of Fig. 2.5, $d_{13} = 2$ and $d_{45} = 1$, as points 4, 5 are adjacent.

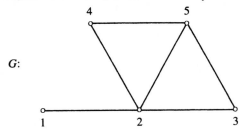

G:

Fig. 2.5. A graph to illustrate walks.

The concept of a maximal set is required in the context where that set is a graph. A set $S = \{x_1, x_2, \ldots, x_n\}$ is *maximal* with respect to some abstract property P if S satisfies P but no set that properly contains S (consists of all the elements of S and has additional elements) does. We will specialize this to the case where S is a graph (regarded as its set of points and lines) and property P is connected.

A graph is *complete* if every pair of points is adjacent and *connected* if every pair of points is joined by a path. A *subgraph* of G is a graph having all its points and lines in G. A maximal connected subgraph of G is called a *connected component* or simply a *component* of G. In Fig. 2.6, the graphs to the right of the dotted line are connected, and those to the left are disconnected. The last graph is complete. The first graph consists of four (trivial) components, and the second has three components.

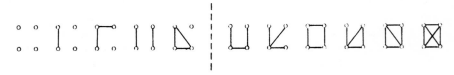

Fig. 2.6. The graphs with four points.

The established notation for the complete graph with p points is K_p, so that the last graph in Fig. 2.6 is K_4. It is easy to see that the number of lines in K_p is $p(p-1)/2$. For example, K_4 has $4 \cdot 3/2 = 6$ lines. A concept that occurs frequently in describing the structure of a graph is its density. The *density* of a (p, q) graph G is the ratio of the number q of lines of G to the number of lines in K_p, that is,

18

$$\frac{q}{p(p-1)/2} = \frac{2q}{p(p-1)} = \frac{\sum d_i}{p(p-1)} .$$

A graph is called *sparse* if its density is low. Note that the density of a graph does not convey structural properties – it does not even tell whether G is connected.

A *cutpoint* of a connected graph is one whose removal (together with its incident lines) disconnects it, that is, divides it into two or more subgraphs that have no lines joining them and hence are not connected to each other. A point v is *isolated* if $deg\ v = 0$, and it is an *endpoint* if $deg\ v = 1$. All the graphs in Figs. 2.1 and 2.2 are trees and contain only cutpoints and endpoints, whereas neither of the graphs in Fig. 2.4 does.

Finally, an *isomorphism* between two graphs $G_1 = (V_1, E_1)$ and $G_2 = (V_2, E_2)$ is a one-to-one correspondence between V_1 and V_2 that preserves adjacency. This is written $G_1 \cong G_2$ or sometimes $G_1 = G_2$. The two graphs in Figs. 1.1 and 1.2, although they appear to be different, are isomorphic. In graph theory whenever a graph is chosen at random, it is precisely this one, hence it is called "the random graph." Fig. 2.6 shows all the graphs with four points. No 2 of these 11 graphs are isomorphic (collectively, they form a *nonisomorphic* set of graphs). Furthermore, no matter which graph with four points anyone will ever draw, it will be isomorphic with one of these.

These preliminary definitions are sufficient to introduce trees and blocks as models of cognitive and social structures.

Trees

It has been suggested elsewhere (Hage 1978) that Puluwatese mnemonics serve not only to organize geographical relations, but also, and perhaps more important, to store and retrieve vast amounts of other kinds of cultural information, including myths, spells, ceremonies, chants, and recitations associated with them, making them "a kind of literature" (Riesenberg 1972:20), rather than merely a set of lists. This interpretation also accounts for the obviously nonempirical properties of many of the geographical schemata. The mnemonics operate as follows.

Each location in a sequence – an island, a reef, or a whale – is a cue for an item of information, a myth or a spell or part of one. Each sequence contains some homogeneous set of items, a set of myths or spells, or a complete version of one. This type of device is not unique to Puluwat, but is well known from the classical literature on rhetoric as the method of loci or the artificial memory described in Cicero's *De oratoria,* in the *Ad Herennium,* and in Quintilian's *Institutio oratoria,* and also used and discussed in later European learning (Yates 1966). The method consists of using a set of locations – for example, places in a building – and associating with each place an image of the item to be remembered. Items are recalled by mentally traveling from place to place and observing each item stored there. Cicero and later scholars believed the method to be effective, and experimental results show it to be dramatically so, improving recall from two to seven times compared to usual methods (Bower 1970a,b). (One may easily verify this by informal experiment.)

The mnemonic effect is enhanced by using vivid, fantastic, or bizarre imagery and by patterning the locations in some way. As an example, in the first graph of Fig. 2.2, called "Aligning the skids" (the skids used to launch a canoe), all the endpoints are islands and all the cutpoints are whales, conveniently located due south of each island and easily recognizable by their successive contrasts in color, size, and morphology. The second whale has a more pointed head than the first one; the third whale has a wider tail; the fourth whale is white; the fifth whale is accompanied by another "only as long as a man's forearm"; and the sixth has only one fluke on each side. The fact that most of the Puluwatese mnemonics, however heterogeneous they may appear, are based on the same underlying type of graph suggests an additional structural reason for their effectiveness. We refer to the concept of linearity, which has application to simple graphs as well as to directed graphs (in which the lines have arrows on them) and signed graphs (in which each line is either positive or negative).

Classic experiments in cognitive psychology, such as Zajonc and Burnstein (1965), Henley, Horsfall, and De Soto (1969), have shown that certain kinds of structures, modeled by graphs, are easier to learn than others. (In the language of Gestalt psychology they have "good form," in the language of structuralism, one might say they are "good to think.") In particular, balanced signed graphs are easier to learn than unbalanced ones (for important material), and asymmetric transitive directed graphs are easier to learn than nontransitive ones. The reason, we think, is that both types of graphs, balanced and transitive, are linear structures. As will be shown in Chapters 3 and 4, balanced signed graphs are linear in that they condense to two (positive) points joined by a (negative) line, whereas unbalanced ones do not, and asymmetric transitive directed graphs are linear in that their points have a partial or complete order. By generalization, one would expect similar considerations to apply to ordinary graphs. Within the class of connected graphs, trees should be easier to learn than nontrees, and within the class of trees, those that are caterpillars should be easier to learn than noncaterpillars, since they are more linear.

Because Puluwatese navigators spend years learning all there is to know in their universe, it is not surprising that they have systematic recourse to mnemonics. Riesenberg counted over 80 of them, and his count may not be complete. Their preferred structural type of mnemonic is one whose under-

Graphs

lying graph is the most linear or the most easily coded of all connected graphs. These concepts may be explained as follows.

A *tree* is a connected graph with no cycles. All 11 trees with seven points are shown in Fig. 2.7. A path with n points is designated P_n. Fig. 2.8 shows P_4 and P_5.

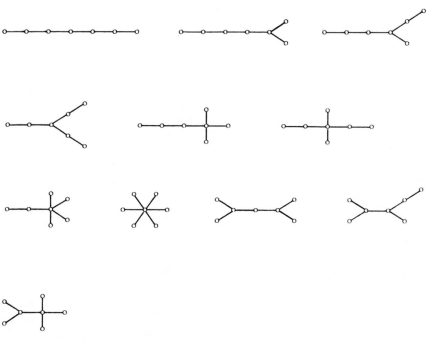

Fig. 2.7. The eleven trees with seven points.

For any tree T, it is customary to write T', called the *derived subtree* for the tree obtained from T by removing every endpoint of T along with its endline. This is shown in Fig. 2.9.

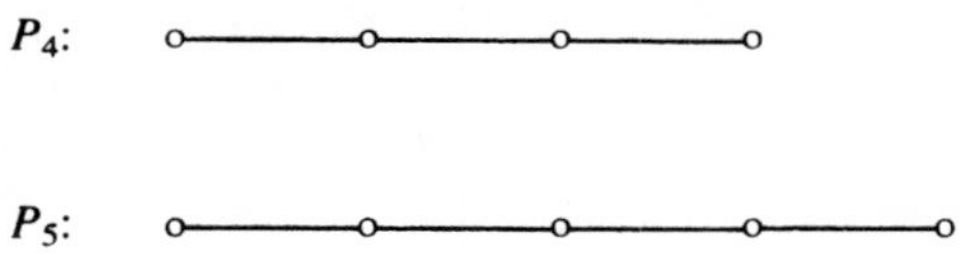

P_4: o———o———o———o

P_5: o———o———o———o———o

Fig. 2.8. Two paths.

A very useful saying among researchers in graph theory recommends that when a new unsolved problem is found to be temporarily intractable for arbitrary graphs, it should be tried for trees. (Examples can be given: Centrality was discovered about 1860 for trees by Cayley, Jordan, and Sylvester, and then about a century later by Harary and Norman 1953, for con-

21

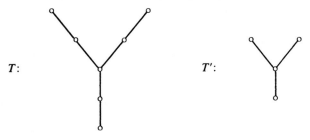

Fig. 2.9. A tree and its derived subtree.

nected graphs in general.) The reason is that among all connected graphs, trees are the simplest to visualize, conceptualize, describe, remember, and characterize. Why is this so? Because a tree is a connected graph with no cycles, it follows at once that in every tree, any two distinct points are joined by a unique path. For as soon as one cycle is present in a graph, each pair of its points is joined by two different paths around the cycle. In this sense, cycles introduce complexity into a structure. (This would explain why the Puluwatese mnemonics probably function more to store and retrieve non-geographical information than to provide geographical orientation. Unlike really adequate spatial orienting schemata, as described, for example, in Lynch 1960, these structures do not allow for alternate routes.) Another reason is that every nontrivial tree is a *minimal connected graph* because it is connected, but the removal of any one of its lines results in a disconnected graph (called a *forest*) with two components.

A similar saying within the theory of trees proclaims that a particular class of trees, called caterpillars, have the simplest structure among all the trees. A *caterpillar* is a tree T with at least three points whose derived tree T' is a path. (The smallest caterpillar is the path P_3 whose derived graph P_3' is trivial, consisting of a single isolated point; that is, the path P_1 containing one point and no lines.) Two caterpillars are shown in Fig. 2.10. Their derived subtrees are precisely the paths P_4 and P_5 of Fig. 2.8. At each point of these two paths in the figure, a number tells the number of endpoints adjacent to it. These two caterpillars are denoted by S and T with $S' = P_4$ and $T' = P_5$. It is clear that all the graphs of the Puluwatese mnemonics shown in Figs. 2.1 and 2.2 are caterpillars and could be drawn in the same way as those in Fig. 2.10.

From a psychological point of view, the relative simplicity of caterpillars can be defined in two different ways. Regarded introspectively, caterpillars are simpler than noncaterpillars because their mental traversal consists of going down a path that can never branch more than one segment away. A noncaterpillar has branches of variable lengths, and so it is easier to get lost in it. One can ignore introspection and adopt the central metaphor of artifi-

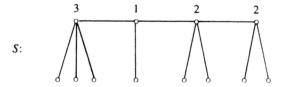

S:

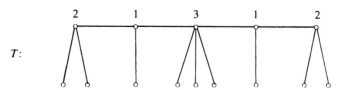

T:

Fig. 2.10. Two caterpillars.

cial intelligence or cognitive science; namely, that mind is program or software. Then the relative simplicity of caterpillars is reflected by the compactness of a code that will identify and generate them.

Let T be a caterpillar such that $T' = P_n$. Let $(a_1, a_2, \ldots, a_n)$ be the sequence in which a_i is the number of endpoints adjacent to point v_i of path P_n. Consider also the backward sequence $(a_n, \ldots, a_2, a_1)$. Call the first sequence *smaller than* the second if $a_1 < a_n$, or if $a_1 = a_n$ and $a_2 < a_{n-1}$, and so on. Define the *code of caterpillar* T as the smaller of those two sequences. Thus in Fig. 2.10, S has the code (2, 2, 1, 3) and T is coded by (2, 1, 3, 1, 2), which is palindromic. Note that the first and last integers in the code of a caterpillar must be positive, but 0 can occur as an inner component in the code vector.

The code of a caterpillar determines it uniquely: Given the code, the caterpillar can be drawn at once. For trees that are not caterpillars, the codification problem is considerably more complicated. Consider, for example, the tree T of Fig. 2.9, which is well known (Harary and Schwenk 1973) as the smallest tree that is not a caterpillar. As shown in Fig. 2.9, its derived tree T' is a star and not a path.

Fig. 2.11 shows this tree with two different arbitrary labelings. Let us trace in Fig. 2.11(a) a closed walk beginning with the endpoint labeled 1:

$$1 \; 2 \; 3 \; 4 \; 5 \; 4 \; 3 \; 6 \; 7 \; 6 \; 3 \; 2 \; 1. \tag{1}$$

Clearly this walk (1) determines not only the structure of the tree T of Fig. 2.9, but also the particular labeling it has in Fig. 2.11(a). It is a walk obtained by applying the Method of Tremaux for solving a labyrinth, as

23

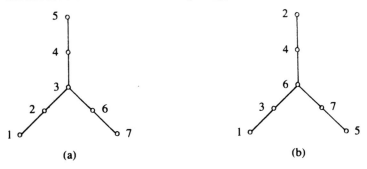

Fig. 2.11. A tree labeled in two ways.

reported in the first book on graph theory (König 1936). However, it is not yet shown to be a code for T, although in fact it is one, as we will now see. The same tree, but with the labeling in Fig. 2.11(b), can be traced similarly by the walk:

$$1\ 3\ 6\ 7\ 5\ 7\ 6\ 4\ 2\ 4\ 6\ 3\ 1. \tag{2}$$

Then the *Tremaux code* of T is the smallest possible such sequence among all the possible labelings of T and all the ways of tracing a closed walk containing each line twice, once in each direction, with smallest defined as for caterpillar codes. It is thus clear that (1) is the smallest possible code for this tree T. Hence (1) is *the* Tremaux code of T.

Unfortunately, there is no similarity between the Tremaux code of a tree that is not a caterpillar and the code of a caterpillar. The Tremaux code of a caterpillar is much longer than the concise caterpillar coding, it contains several repetitions, and it necessarily yields a labeled caterpillar rather than characterizing its basic unlabeled structure.

An application of graph theory to an apparently heterogeneous set of primitive mnemonics provides a characterization of their common underlying structure, and an explanation of this commonality leads to a generalization of a psychological proposition. Thus, graph theoretic analyses of real-life schemata may have implications for cognitive psychology as well as for cultural anthropology. Indeed, this is an area in which there may be productive interaction between these two fields, for as Neisser (1976) has emphasized, virtually nothing is known about the contexts, development, and uses of memory in the real world, a deficiency that obviously precludes the formulation of a general theory.

Trees underlie many social and cultural structures. Genealogical diagrams, such as Conklin (1964) and semantic keys, such as Kay (1966) and Tyler (1969) come readily to mind, as do depictions of social hierarchies, such as the oak trees and pine trees Henry (1954) distinguishes in his study of the social structure of a mental hospital. They also occur in sociolinguistics, where they are included under schemata known as "tours," preferred structures for recalling certain kinds of spatial layouts (Linde and Labov 1975). A spectacular example of a tree in the realm of material culture is the Inca *quipu,* a complex recording device in the form of a string figure. In their recent monograph, Ascher and Ascher (1981) remark, in a discussion of implicit and explicit mathematics, that an Inca *quipu* maker and the English mathematician Arthur Cayley, who discovered trees in the nineteenth century, might have had much to talk about (but in which language?).

As an example from the physical sciences, Cayley was asked by a chemist to derive a formula for the number of saturated hydrocarbons $C_n H_{2n+2}$ (in which two linked carbon atoms have just one chemical bond). Cayley realized at once that a structural model for a saturated hydrocarbon is a tree T in which every point has degree 1 (for a hydrogen atom) or 4 (for carbon). He also noticed that the derived tree T' is the carbon subtree of the hydrocarbon tree.[2] The caterpillars among the hydrocarbons have special chemical properties and are those whose carbon subtree is the path P_n. Cayley first simplified the problem to one he could solve, and proved easily that the number of labeled trees with n points is n^{n-2}. (*Motto:* If you can't count a class of trees, then first count the labeled ones.) He then gracefully counted *rooted trees,* in which one of the points is called the *root* and is distinguished from the others (these are structural models for monosubstituted hydrocarbons in which one of the hydrogen atoms is replaced by a halogen, say chlorine). With difficulty, Cayley counted the unrooted trees and finally succeeded in resolving the original problem of enumerating hydrocarbon trees (see Cayley 1891).

There is a well-known anthropological parallel to Cayley's work, A. Weil's (1969) algebraic interpretation of marriage laws, which appeared in *Les structures élémentaires de la parenté.* Weil used group theory to aid in the study and classification of marriage laws and thereby virtually created a new subfield of anthropology. Considering Lévi-Strauss's general emphasis on the constraints on mental structures and his frequent evocation of the periodic table as a model of structural analysis, it occurs to us, although we cannot pursue it here, that graphical enumeration, a well-developed branch of the combinatorial aspect of graph theory (Harary and Palmer 1973) may provide interesting results in the anticipation and explanation of anthropological structures.

[2] A hydrocarbon has an energy that varies inversely with the total distance of the hydrogen atoms from one another. It is also true that those hydrocarbons with the smaller energy occur more frequently in nature. Clearly, the hydrocarbon in which the n carbon atoms form a carbon tree that is the path P_n will have their hydrogen atoms (the endpoints of the hydrocarbon tree) at maximum total distance. This is precisely the caterpillar hydrocarbon, and it occurs most often in nature among all the isomers of $C_n H_{2n+2}$.

In a tree, there are no cycles. A radically different type of graph is a block, in which every pair of points lies on a common cycle.

Blocks

Erik Schwimmer (1973, 1974) describes a system of gift giving in a Papuan village consisting of 22 households. The gift consists of cooked taro presented by women but conceived of as mediating relations between households. The taro gifts express intimacy and mutual support in economic and political affairs. A significant feature of this system is the making of requests using taro partners as mediators: A asks B to ask C, rather than asking C directly. According to Schwimmer:

The Orokaiva are wary of relations with persons with whom they are not intimate. They do not like to make even simple requests of such persons but prefer to make the request through intermediaries who are intimate with both sides. In that case it is possible, if the request cannot be met, for both sides to pretend that it was never made. No hard words are spoken, nobody is overtly humiliated. There is a customary blank, expressionless style which is used by intermediaries for the transmitting of requests – symbolic gestures of neutrality. The mediator avoids the risk of endangering his own alliance while still making what effort he can to satisfy the party seeking the benefit. If he is successful, the beneficiary is to some extent indebted to him. (Schwimmer 1974:232)

It appears the mediating chains may be of considerable length:

The procedure followed is that A and B use one or more persons to mediate between them. In the simplest case, where A and B are separated by only one link in the chain, they share the same intimate associate who can act as mediator. If they do not share an intimate, i.e., if they are separated by two or more links in the chain, an accommodation may be reached through a series of mediators who act vicariously on behalf of the principal parties. (Schwimmer 1973:134)

Schwimmer describes this system with explicit reference to graph theory. He represents it by a graph similar to the one shown in Fig. 2.12, where the points represent households and the lines represent taro exchange partners. His graph analysis shows two structural features: the "cluster," defined as "intimate one-degree links between all the members" that would correspond to a maximal complete subgraph of G (sometimes called a "clique"), and the "circuit," which unites households through intermediaries and would correspond to a cycle in G.

To understand this description, we require another property of graphs. A *spanning subgraph* of G is a subgraph containing all the points of G. If a graph G has a spanning cycle Z, then G is called a *hamiltonian graph* and Z is a *hamiltonian cycle*. In Fig. 2.13 G_1 is hamiltonian, whereas G_2 is not.

Schwimmer notes that the "circuit is close to what graph theorists call a

Graphs

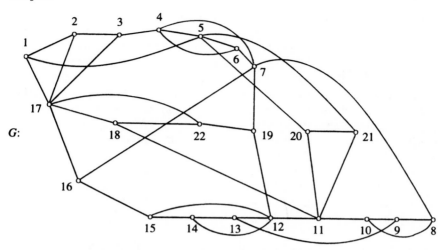

Fig. 2.12. The graph G of the Orokaiva taro exchange system.

Fig. 2.13. A hamiltonian and a nonhamiltonian graph.

hamilton circuit . . . and would be consistent with an almost total absence of stratification'' (1974:231). He enumerates two such circuits, one which connects 19 of the 22 households and another which connects 11 households of one clan formation. Although neither of these is actually hamiltonian, Schwimmer's remarks about the association of such cycles with structural unity and a relative absence of stratification raise a general question about the significance of cycles, hamiltonian or not, in graphs of communication structure.

Actually, it may be difficult to determine in any particular case whether a graph is hamiltonian, because there is no generally usable criterion and the consensus among computer scientists is that no such criterion can be found. There are necessary conditions – for example, G must be connected, contain at least three points, and not contain a cutpoint – and there are sufficient conditions – for example, G is hamiltonian if it is complete or less stringently if it contains enough lines, as in Dirac's theorem (Harary 1969:68), whose hypothesis is that each point has degree at least $p/2$. But a graph consisting of a single cycle is hamiltonian and has fewer lines (only p). Aside from the formal difficulty in characterizing a graph as hamiltonian – the

27

graph of the taro exchange system for example – there is a more important practical consideration. For many graphs of communication or exchange systems, it may be irrelevant that every point lie on a spanning cycle of G, but highly important that, less restrictively, every pair of points lie on a common cycle of G; that is, there is a cycle of G containing both of them.

The latter condition will be seen in Theorem 2.1 as necessary and sufficient for a graph to be a block. By definition, a *block B* is a nontrivial connected graph containing no cutpoints. We see from Fig. 2.6 that there are just three blocks with $p = 4$. All hamiltonian graphs are blocks, but not conversely. In Fig. 2.13 both G_1 and G_2 are blocks. If G is a block, then certain conclusions can be drawn about the integrity, the flow of information, and the relative stratification of the system insofar as this is structurally determined. These conclusions result from an application of the following theorem, given in Harary (1969) and applied by Hage (1979b).

Theorem 2.1. Let G be a connected graph with at least three points. The following statements are equivalent:
(1) G is a block.
(2) Every two points of G lie on a common cycle.
(3) Every point and line of G lie on a common cycle.
(4) Every two lines of G lie on a common cycle.
(5) Given two points and one line of G, there is a path joining the points that contains the line.
(6) For every three distinct points of G, there is a path joining any two of them that contains the third.
(7) For every three distinct points of G, there is a path joining any two of them that does not contain the third.

The graph G of the taro exchange system is a block. Therefore we know by condition (1) that there is no household whose disappearance or demise will disrupt communication within the group (by disconnecting its graph), and by condition (4) that if any one household loses or falls out with a preferential partner, any two households will still be able to communicate. By condition (2), every pair of households has (at least) two alternative paths by which to make requests of each other. Further, by condition (5), any particular link may occur in a mediational chain between any two households, and by conditions (6) and (7), any particular household may occur in a chain, but no particular household must necessarily occur. The communication structure is therefore invulnerable and flexible. And it is unstratified in the sense that no individual or social dyad is able to control communication between any two households or sets of households.

One of the origins of social network studies in anthropology was the study of societies lacking traditional named social groups, in particular those based on kinship (Mitchell 1969). By studying networks, anthropologists were able to determine other ways in which social behavior is regulated or influenced. One advantage of graph theory is that it allows for the determination and characterization of various types of implicit or submerged network structures. Blocks may turn out to be a common type of group in primitive as well as modern societies. They permit communication that is flexible, through the use of alternative channels, and also redundant, through the use of multiple channels – an important consideration in bringing pressure to bear in norm enforcement. They preserve some of the integrative properties of a complete graph and yet require as few as p relations (a single cycle) – an important consideration in the case of large or expanding groups where the number of potential social relations is limited by the amount of material at the disposal of each member to transact, for example, the amount of taro. Conversely, of course, G may have as many as $(p-1)(p-2)/2+1$ lines and still not be a block; that is, G consists of a $p-1$ clique together with one point joined to it by a single line, in which case it has two blocks.

The structure just analyzed consists of a single block. In some cases, the graph of a social group may be connected but consist of a number of blocks, each pair of which is joined by a cutpoint or by a bridge. A *bridge* of G is a line whose removal disconnects a graph, as illustrated by B_2 in Fig. 2.14. Each bridge of a graph is itself a block. In such a case, the individuals and relations represented by cutpoints and bridges would presumably be of fundamental importance in the group.

We note also that our example involves only symmetric relations. Blocks need not be so restricted in application. Blocks are applicable to nonsymmetric relations in directed graphs (Harary, Norman, and Cartwright 1965)

G: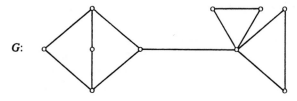

Fig. 2.14. A graph and its blocks.

and to antithetical relations in signed graphs. For example, the New Guinea alliance system shown in the next chapter is a block. It appears to be typical of Highland New Guinea, where societies are sometimes defined as "zones of interaction" (Berndt 1964).

So far we have considered trees, graphs in which every pair of points is joined by a unique path, and blocks, graphs in which every pair of points is joined by multiple paths. We now turn from path structure *per se* to path length or distance in graphs, with special reference to the concept of centrality.

Centrality

There are two classic concepts of centrality for a connected graph G, and in particular for a tree T, called the center and the centroid. In order to define the first of these, discovered by C. Jordan (1869), it is convenient to say that the *eccentricity* of a point v of G is the maximum distance between v and any other point. Then the *center* $C(G)$ consists of all points of smallest eccentricity. Jordan proved that the center of any tree T consists of just one point or of two adjacent points, as illustrated in Fig. 2.15.

Fig. 2.15. Two trees with central points displayed prominently.

J. J. Sylvester (1882) introduced the centroid of a tree using the concept of the weight of a branch. For a given tree T having a point v incident with a line e, the *branch* $B(v, e)$ consists of v and e and all points and lines reachable by a path from v that begins with line e. Thus $B(v, e)$ is a subtree of T and the *weight of this branch* is the number of lines in it. The *weight at point v* is the maximum weight of any branch at v. Then the *centroid* of T consists of all points with smallest weight. Sylvester proved that the centroid of any tree T consists of just one point or of two adjacent points. Fig. 2.16 shows two trees, the first with two central points and one centroid, and the second oppositely endowed.

These two centrality concepts are useful in graphical enumeration and in the mainstream of graph theory. More recently, many alternative notions of

(a)   (b)

Fig. 2.16. Two trees with central points circled and centroid points boxed.

centrality have been introduced, motivated by their potential applicability to the social sciences.

Intuitively, certain individuals, groups, or communities in a social system may be "more central" than others and therefore more prominent, influential, or powerful. The sociologist Linton Freeman (1977, 1979), in his review of the literature on experimental small groups, distinguishes three quite distinct graph theoretic meanings of centrality.

The first, the *degree of a point,* refers to the number of lines incident with a point and is interpreted as an index of "communication activity." In this metaphor, the higher the degree of a point, the more central it is considered to be. An anthropological example would be the egocentric networks familiar from social network studies, such as Barnes (1969, 1972) and Mitchell (1969). These structures are *stars,* trees consisting of one cutpoint and $p - 1$ endpoints. Stars overlap in a connected graph, as illustrated in Fig. 2.17.

Centrality models are obviously not limited in application to the study of small groups, as may be shown by an archeology example that uses degree size and eccentricity in graphs. In Mayan studies, Norman Hammond (1972a,b) has proposed that

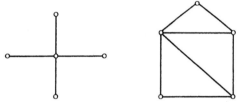

Fig. 2.17. A star and a graph containing five stars.

ceremonial precinct layouts are due to deliberate planning and not to environmental contingency, as the prevailing view would have it. In his analysis of the Classic Mayan ceremonial complex of Lubaantún in British Honduras, Hammond finds an apparently concentric pattern in the layout of functionally differentiated types of plazas: A large religious plaza is in the center, with residential plazas on the periphery and ceremonial plazas in between. He hypothesizes that such a concentric zone model should have a "traffic plan" such that public areas (religious and ceremonial plazas) would be most accessible and private areas (residential plazas) least accessible. By accessibility, Hammond means degree size and eccentricity. Fig. 2.18 from Hammond (1972a) shows the plazas classified by type and manner in which they are connected. It is clear that in terms of degree size, the major religious plaza (IV) is most central in this graph, whereas residential plazas are least central, with ceremonial plazas most often in between.

Fig. 2.19 is another version of the same graph, but this time showing the eccentricity of each point. The center of G is again plaza IV, because it has minimum eccentricity (3), whereas the periphery of G, points with maximum eccentricity (6), consists of three of the residential plazas, the other three having eccentricity 5. The ceremonial centers have either 5 or 4.

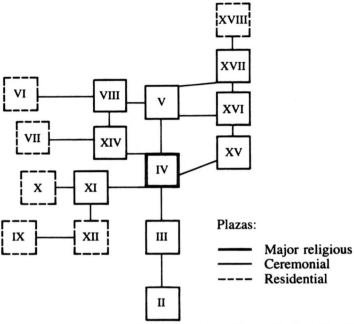

Plazas:

——— Major religious
——— Ceremonial
- - - Residential

Fig. 2.18. A graph of plazas in a Mayan ceremonial complex (from Hammond 1972a).

By either measure, then, public areas are more accessible (central) than private ones.[3]

Hammond is able to draw certain other conclusions from the application of this model, including an informed guess as to which plaza served as the marketplace (plaza VIII) and a conjecture that because the earliest ball court was located near an area of restricted access (between plazas II and III), the famous Mayan game of *pok-ta-pok* was originally "confined to an elite group within a well-stratified society" (Hammond 1972b:89). He also speculates on broader archeological applications of such models:

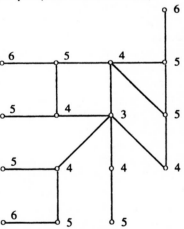

Fig. 2.19. The eccentricities of the points in the graph of Fig. 2.18.

[3] A note on Hammond's special terminology is required. He uses "accessibility" generically to include both "accessibility," the degree of a point, and "centrality," the eccentricity of a point. In order to combine these two measures by summation so that the higher the number, the greater the accessibility, he subtracts the eccentricity of a point from the diameter of the graph.

32

Graphs

Analysis of the centrality and accessibility of the different areas that make up "palace" complexes in, for example, Mesopotamia or Crete or Mycenaean Greece might suggest functions quite different from those enshrined in long accepted but essentially poetic phrases such as "the queen's antechamber" or "lustral area." (Hammond 1972b:89)

The second definition of centrality, *distance* or *closeness,* refers to the sum of the lengths of the shortest paths (geodesics) between one point and all other points of a connected graph, and is interpreted as an index of "communication efficiency." In the graph in Fig. 2.20, the total distance from each point to all other points is shown in parens. Points 1 and 2 are most central, or closest, and points 4 and 5 are most distant, or peripheral. Examples from economic anthropology and archeology would be the relative locational advantage of communities in trade networks as analyzed by Brookfield and Hart (1971) and Irwin (1974, 1978) in two Melanesian systems.

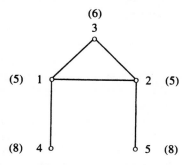

Fig. 2.20. A graph illustrating total distance.

As a sociological example of this definition of centrality, one might cite (following Barnes and Harary 1983) W. E. Armstrong's (1928) concept of "tribal distance" as a model for the characterization of social groupings as opposed to discrete social groups in the analysis of Melanesian social structure. A grouping is

best regarded as a "sequence" of persons which may or may not have a definite end. A given grouping of A is, therefore, the sequence of persons starting from A, arranged in order of "closeness" to A. (Armstrong 1928:33)

A grouping may thus be regarded as a "sequence" somewhat in the sense used in mathematics. The remoter the term in the sequence, the less does that term have the defining quality of the grouping. To say that a family is a grouping is to say that an indefinite number of persons can be arranged in a sequence relative to a given individual, and that the position in the sequence defines the family distance of that person from the given individual. (Armstrong 1928:33)

The sequence, or as we should say, the path, is defined by a variety of social factors to be empirically determined in each ethnographic case.

The third definition, called betweenness, is less obvious than the first two, but potentially of greater anthropological significance if one considers how fundamental the concept of mediation is in exchange and communication networks. *Betweenness* refers to the frequency of occurrence of each point on the geodesics between all pairs of other points and is an index of

33

Structural models in anthropology

the "potential for control of communication." In effect, this is a generalization of the notion of a cutpoint.

A reading of Melanesian ethnography suggests that this third concept of centrality may serve to elucidate the nature of leadership in informal big man systems.[4] These are systems in which the status of individuals, who in the vernacular are commonly called "big men," is achieved largely through the manipulation of exchange relations (Sahlins 1963) or better, through the manipulation of exchange networks (Strathern 1971). The application of such models has particular urgency because as Forge (1972) has pointed out, in spite of Malinowski's (1961) ethnography and Mauss's (1954) analysis, and in spite of the salience of all manner of exchange, ceremonial and otherwise, in Melanesian societies, there has been little in the way of detailed analysis of particular exchange systems and little appreciation of their underlying structural properties. To illustrate the model, we will apply it to the Orokaiva taro exchange system described above.[5] In explaining its structure, we will first provide a purely intuitive example.

In the graph G of Fig. 2.21, consider all the shortest paths between the points 1 and 4. There are three such paths, 1 2 3 4, 1 2 5 4, and 1 6 5 4. Notice that points 2 and 5 lie on two of the three shortest paths joining 1 and 4. Therefore, one can intuitively regard points 2 and 5 as more central than points 3 and 6, which lie on just one of these three shortest paths. For each pair of points in the graph, we now enumerate all the shortest paths having two or more lines. Then the points that lie on the highest proportion of these paths are relatively more central than the others.

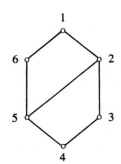

Fig. 2.21. A graph illustrating betweenness.

Such points, representing, say, individuals or groups, clearly have greater potential for the control of communication in a system. We now state the necessary formal definitions.

Following Freeman (1979), let g_{ij} be the number of $i-j$ geodesics and let $g_{ij}(v_k)$ be the number of these geodesics containing v_k. Then he defines the *betweenness value of v_k with respect to the point pair v_iv_j as the ratio,*

$$b_{ij}(v_k) = \frac{g_{ij}(v_k)}{g_{ij}} ,$$

which is of course the probability that a randomly selected $i-j$ geodesic contains point k. He then defines in terms of these probabilities the *partial*

[4]For an ideal type characterization of such systems, see Sahlins (1963).
[5]For a more extensive description and further suggested applications of the model, see Hage and Harary (1981a).

34

Graphs

betweenness value of v_k (independent of any one pair of points) by the formula

$$C_B(v_k) = \sum_{j=1}^{P} \sum_{i=1}^{j-1} b_{ij}(v_k).$$

He comments that whenever v_k lies on every $i-j$ geodesic, the point pair i,j contributes 1 to the sum $C_B(v_k)$. The particular case that there is a unique geodesic joining each pair of distinct points occurs only when G is a tree (a condition which, as we have seen, serves to define a tree).[6] When there are alternative geodesics, $C_B(v_k)$ grows in proportion to the frequency of occurrence of v_k among these alternatives. It can easily be imagined that locating geodesics and counting become difficult for large graphs. In Chapter 5, we show how this may be done using matrix methods.

Freeman mentions the immediate fact that among all p point graphs, the maximum possible value of $C_B(v_k)$ is $(p-1)(p-2)/2 = (p^2 - 3p + 2)/2$, as illustrated for $p = 5$ in Fig. 2.17 by the 5-point star. He thus defines the *relative centrality* $C_B'(v_k)$ as the ratio of $C_B(v_k)$ to this maximum expression, so that

$$C_B'(v_k) = \frac{2C_B(v_k)}{p^2 - 3p + 2}.$$

The unique point (if any) with the highest value of $C_B'(v_k)$ is called the *most central point* of the graph in this sense. Freeman defines the centrality of a connected graph as "the average difference between the relative centrality of the most central point, $C_B'(v^*)$, and that of all other points," giving:

$$C_B(G) = \frac{\sum_{i=1}^{P} [C_B'(v^*) - C_B'(v_i)]}{p - 1}.$$

Note that $C_B(v_k)$ measures the centrality of a single point, whereas $C_B(G)$ measures that of the entire graph.

To illustrate these measures, Fig. 2.22 shows a graph G together with its centrality $C_B(G)$, and the partial betweenness value $C_B(v_k)$ and the relative centrality $C_B'(v_k)$ of each of its points. As a further illustration of graph centrality, when G is a star, $C_B(G) = 1$, and when G is complete, $C_B(G) = 0$.

When the measures of degree centrality, distance centrality, and betweenness centrality are computed, there may not be a perfect correlation among them. In Fig. 2.22, for example, points 2, 4, 5 all have degree 2, but points 4 and 5 are more between than point 2. More important, as Freeman emphasizes, the choice of a definition of centrality should be determined by the particular structure being studied.

---

[6]Then Freeman's sum $C_B(v_k)$ becomes precisely the cutting number of a point v_k of a tree as studied by Harary and Ostrand (1971). (By definition, the *cutting number* of a point v_k of a connected graph is the number of pairs of points u and w such that every $u-w$ path contains v_k as an intermediate point.)

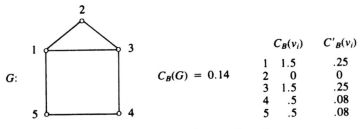

Fig. 2.22. The measures of betweenness in a graph.

In some situations, all three definitions or some combination of them may be appropriate. It may be remarked in this connection that the geographer F. R. Pitts's (1965) distance analysis of centrality in a network of Russian medieval towns, which was the model for the anthropological analysis of a Melanesian system (Irwin 1974, 1978), was subsequently reevaluated (Pitts 1979) using both distance, to measure system effort required for travel, and betweenness, which would reflect the characterization of the most prominent communities as "at the crossroads" between others. (The problem of using the appropriate graph theoretic model will come up later in a discussion of signed graphs.)

Consider now the Orokaiva exchange system represented by the graph in Fig. 2.13. Schwimmer describes it as "unstratified." There are, however, leaders, "big men" in the village. As a result of operating as a mediator in the transmission of requests, an individual puts the parties involved under some obligation for general aid and support in intravillage affairs. One would therefore expect leaders to rank high on betweenness.

Table 2.1 shows the betweenness measures for all 22 households and the centrality of the entire graph. The table shows the households, together with their clan membership. While $C_B(v_k)$ gives the unique partial betweenness value of each household, $C_B'(v_k)$ gives the relative betweenness of each. The latter disregards the size of the network and can be used to compare graphs of different size. The final column gives the rank of each household in relation to all the others. Then the centrality $C_B = C_B(G)$ of the entire system is 0.29. This value indicates a relatively egalitarian system. It should be a matter of considerable theoretical interest to compare the values of C_B for an entire set of systems and then proceed to examine their infrastructural correlates.

The acknowledged village leader, 12, ranks second within his own clan, Seho, and fourth within the village. Since his form of leadership is described as an "aristocratic type of New Guinea 'big man' leadership" (Schwimmer 1973:133) that utilizes the power of intimate links (graph theoretically, his *neighborhood,* the set of points to which he is adjacent), his power is even greater because his neighborhood includes 11, who ranks first villagewide,

Graphs

Table 2.1. *Centrality of households in the taro exchange system*

Household	Clan	$C_B(v_k)$	$C_B'(v_k)$	Rank
1	Sorovi	8.2667	0.0394	13
2	Sorovi	0.5000	0.0024	22
3	Sorovi	5.8333	0.0278	15
4	Sorovi	15.4333	0.0735	8
5	Jegase	23.6000	0.1124	5
6	Jegase	4.5167	0.0215	20
7	Jegase	41.4500	0.1974	2
8	Jegase	15.1667	0.0722	10
9	Seho	5.5000	0.0262	19
10	Seho	11.0000	0.0524	12
11	Seho	46.3833	0.2209	1
12	Seho	34.1333	0.1625	4
13	Seho	5.6667	0.0270	16
14	Seho	2.3333	0.0111	21
15	Sorovi	12.1833	0.0580	11
16	Seho	23.5500	0.1121	6
17	Sorovi	38.1833	0.1818	3
18	Jegase	15.3500	0.0731	9
19	Jegase	17.5333	0.0835	7
20	Jegase	5.6500	0.0269	17
21	Jegase	5.6500	0.0269	17
22	Jegase	7.1167	0.0339	14

and also 19, who ranks seventh. The challenger to 12 is 5, who ranks second within his own clan, Jegase, or third if Jegase and Sorovi are considered as a single clan formation, and fifth villagewide. In a system such as this, mediators can be considered not only in terms of power, but also in terms of their contribution to the integration of a group through their facilitation of communication. Of the individuals who "had peaceful natures and tried to keep on friendly terms with almost everyone," 17 and 16 rank third and sixth, respectively.

The model serves to characterize the system rather well in predicting the location of leadership. We must, however, hasten to point out that it makes two assumptions that could be modified in the light of additional ethnographic data, and one of which undoubtedly would have to be modified for some (but not necessarily for all) real-life networks.

The first is that communication occurs along geodesics. In some instances it does occur along paths of greater length, but it may not be unreasonable to suppose that in general, as long as communication is indirect, geodesics would be more likely to be used because less effort is required and because there is less decay or distortion of the message the fewer the links it traverses. Schwimmer's description of the simplest case suggests that geodesics are often used.

The second is that the choice of geodesics is random. Although this is not

37

an empirically correct assumption, it serves the purpose of providing a first approximation justified by the information available. With additional information, a subsequent modification of the random graph model, in which the links are weighted, could be proposed. Schwimmer, for example, discriminates first, second, and third preferential partners on the basis of the frequency of taro transactions (although in many cases the differences between these categories are negligible). The new model would be a tree in which probabilities are multiplied over successive lines in a path.

The betweenness model of centrality potentially has broad anthropological application. In Melanesian studies, it appears to capture an essential aspect of the big man form of leadership. Concerning one of the best known of Highland New Guinea societies, A. Strathern[7] remarks that Mt. Hagen big men definitely see themselves as communication mediators. The model is easily generalized to other kinds of systems – for example, to trade networks as noted above, and to kinship systems such as the Navaho, in which an essential feature of status is defined by the indirect as well as the direct making of requests (Aberle 1961). There are also other correlates of mediational activity. A psychodynamic one is Shimbel's (cited in Freeman 1977):

> Suppose that in order for site i to contact site j, site k must be used as an intermediate station. Site k in such a network has a certain "responsibility" to sites i and j.
> If we count all of the minimum paths which pass through site k, then we have a measure of the "stress" which site k must undergo during the activity of the network. A vector giving this number for each member of the network would give us a good idea of stress conditions throughout the system. (Shimbel 1953:501)

Finally, with regard to fieldwork, the model can serve both to validate ethnographic intuition and informants' statements concerning the locus and structure of leadership and also to elucidate the same in informal systems.

This analysis, which was concerned with the pattern rather than with the number of relationships in an exchange system, suggests a general observation concerning the relation between structure and density in graphs. In the anthropological literature on social networks, there has been much preoccupation strictly with the number of lines in a graph and the number of lines incident with each point (its degree). The degree sequence, however, is a weak indicator of structure.

The *degree sequence* of a graph is a list of the degrees of all the points; conventionally, this begins with the maximum degree and continues in nonincreasing order. It is well known in graph theory that the smallest number of points in two nonisomorphic graphs having the same degree sequence is five. The example is given by the two different graphs G_1 and G_2 of Fig. 2.23, both of which have degree sequence 3, 2, 2, 2, 1; this can be written more briefly as $3\ 2^3\ 1$. Using this concise notation, the degree

[7]Personal communication.

38

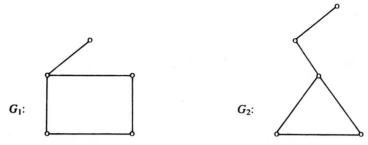

G_1: G_2:

Fig. 2.23. Two nonisomorphic graphs having the same degree sequence.

sequence of the taro exchange system of Fig. 2.12 is 6 5^3 4 3^{17}. There is a large number of graphs having this degree sequence.

An observation of J. C. Mitchell's, which can be taken generically, expresses the matter well:

> In sociological analysis our interest is primarily in reachability since norm enforcement may occur through transmission of opinions and attitudes along the links of a network. A dense network may imply that this enforcement is more likely to take place than a sparse one but this cannot be taken for granted. The "pattern" of the network must also be taken into consideration. (Mitchell 1969:18–19)

Graph theoretic models are applicable to the analysis of both social relations and cognitive schematas. By finding the type of graph underlying an empirical structure, we can draw certain conclusions about it. The discussion so far has been primarily concerned with path structure in certain classes of graphs. We now turn from a consideration of reachability and distance in graphs based on symmetric relations to the partitioning of points in graphs based on antithetical relations.

3

Signed graphs

The human mind arranges phenomena by a constant process of cognition, identification and polarization. Calling a certain object black presupposes both an identification of this object with other objects called black and a contra-distinction of this object to other objects called white. Things do not have absolute self-sustaining properties. The latter are always a matter of *relative position*.

Jan Pouwer, "Towards a configurational approach
to society and culture in New Guinea"

New Guinea alliance systems have been characterized by such notions as "fields of warfare" (A. Strathern 1971), "zones of interaction" (Berndt 1964), and "clusters" of clans (Ryan 1959). In his analysis of the structure of feasting and warfare among the Polopa of Highland New Guinea, D. J. J. Brown (1979) uses a metaphor based on a relation of coloring to describe the intertribal pattern of friendship and enmity:

In Polopa conception, the enemies of enemies (and allies of allies) are always allies, by definition. When visiting communities unknown to my companions (in making a survey of the Polopa area), their first concern was to establish whether or not they had allies or enemies in common, and if not they prevailed on me to move on. They did not recognize the possibility of neutrality. Some results of the survey are shown in the accompanying map ... of the communities of Polopa and their immediate neighbors. For convenience the communities are color-coded, marking the position of each by an R (red) or G (green): any two of the same color are allies and of different colors enemies. Notice that two colors only are sufficient to complete the pattern. It has a patchwork quality, but of red and green only. This means that the enemies of enemies are indeed allies: the pattern is a transitive one. (Brown 1979:727)

This characterization does not correspond to the standard definition of a transitive relation (see Chapter 4), but it does correspond to a balanced signed graph, which is complete because there are no neutrals.

Signed graphs provide models for the analysis of antithetical relations such as those based on the oppositions, friend versus enemy and intimacy versus restraint. Our presentation of these models focuses on the theory of structural balance and more generally of clustering, as exemplified in systems of kinship and political alliance. Kinship applications are noteworthy not only because they illustrate native as well as observer models of attitudinal consistency, but also because they raise the problem of how the symbols + and −, which figure so prominently in structuralist analysis, are to be interpreted, and how particular types of graphs are to be coordinated to different empirical structures. The most interesting applications to political anthropology would be diachronic in nature. Where historical data are lacking, it is often possible to gain some insight into dynamic processes by means of simulations such as the one to be described here.

Two or more points of a graph are called *independent* if they are pairwise nonadjacent. When this is so, these points can be assigned one color, a different set of independent points can have another color, and so forth. The resulting partition of the point set of the given graph or signed graph is thus based on the coloring. Balance and clustering phenomena in a signed graph are also based on partitions of its set of points. In order to acknowledge historical precedence within graph theory, and because we think both types of models are applicable to the analysis of equivalence structures in anthropology, we begin by defining a coloring of a graph.[1]

Coloring of a graph

Historically, the study of graph coloring was motivated by the coloring of geographical maps in which two countries (or other distinct regions) with a common boundary must be given different colors in order to identify their territories unambiguously. When each country is replaced by one point (standing for its capital city), two of these points are joined by a line whenever their countries have a common boundary. The resulting graph, called the *dual* of the map, suggests the next definition. A *coloring of a graph* is an assignment of colors to its points so that no two adjacent points have the same color. The set of all points with any one color is independent and is called a *color class*. An *n-coloring* of a graph G uses exactly n colors; it thereby partitions V into n color classes. The *chromatic number* $\chi(G)$ is defined as the minimum n for which G has an n-coloring. A graph G is called *n-colorable if* $\chi(G) \leq n$ and is *n-chromatic* if $\chi(G) = n$.

Since G obviously has a p-coloring and a $\chi(G)$-coloring, it must also have an n-coloring whenever $\chi(G) < n < p$. The graph of Fig. 3.1 is 2-chromatic; n-colorings for $n = 2, 3, 4$ are displayed, with positive integers designating the colors.

The chromatic numbers for some of the familiar graphs are easily

[1]The coloring of a graph is discussed in the Appendix because it turns out to be the basis for the notion of a homomorphism, which has been proposed as a model for the simplification of complex social and cultural structures.

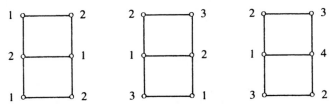

Fig. 3.1. Three colorings of a graph.

determined, such as those for a complete graph $\chi\,(K_p) = p$, and for any nontrivial tree, T, $\chi\,(T) = 2$. A graph G is called *planar* if G can be drawn in the plane so that no two of its edges cross each other. It has long been conjectured and recently proved (Appel and Haken 1976; see also Kainen and Saaty 1977) that every planar graph is 4-colorable.

There is a simple anthropological example of the coloring of a graph. In his monograph *Naven*, Gregory Bateson (1958) isolates a structural principle of Iatmul *eidos*, which he calls "diagonal duality." This principle is expressed in the kinship system in the merging of alternate generations; in the equivalence of ego and his father's father, as opposed to ego's son and father, and so on. The graph of this structure is the path P_n in which the two different colors correspond to adjacent generations and the two color classes correspond to the merged alternate generations. The same type of tree, a path, also underlies Iatmul musical performance, in which two flutes, each of which has alternate notes of a scale and both of which make up a single order, are played together. And it underlies Iatmul initiation, in which adjacent age grades are in different moieties. Since Bateson's geometrical analogy of diagonal duality is the classification of the opposite corners of a rhombus as equivalent, one could define his principle generally as a 2-colorable graph.

Coloring of a signed graph

A *signed graph S* is obtained from a graph G when each line of G is designated either positive or negative. An *n-coloring of a signed graph S* is an assignment of colors to its points such that (1) every two points joined by a negative line are in different color classes, and (2) every two points joined by a positive line are in the same color class. We say that *S has a coloring* or is *colorable* if it has an *n*-coloring for some *n*. It follows immediately from these definitions that if a signed graph *S* has only negative lines, the problem of coloring *S* is the same as that of coloring its underlying unsigned graph. If, however, *S* has some positive lines, it may not be colorable.

Let S^+ be the spanning subgraph obtained by removing all negative lines from S. Recall that a component of a graph is a maximal connected subgraph. The *positive components* of S are just the components of S^+. It follows from this definition that two distinct points of S are in the same positive component if and only if they are joined by a path consisting entirely of positive lines (called an *all-positive path*). Clearly, the positive components of S partition $V(S)$ into subsets such that each positive line joins two points in the same subset, and S has exactly one such partitioning.

The following theorem (Cartwright and Harary 1968) presents a criterion for a signed graph to have a coloring.

Theorem 3.1. A signed graph S has a coloring if and only if S has no cycle with exactly one negative line.

This theorem is illustrated in Fig. 3.2, which shows a colorable signed graph S (in which negative lines are represented by dashes). Its three positive components are evident in S^+, and its point set $V(S)$ can be partitioned into the color classes $\{v_1, v_2\}$, $\{v_3\}$, $\{v_4, v_5\}$. Clearly, S has no negative lines joining two points of the same positive component, nor does it have a cycle with exactly one negative line.

The *condensation of S by its positive components,* denoted S^*, is the signed graph whose points are the subsets $\pi_1, \pi_2, \ldots, \pi_n$ determined by (contracting or shrinking to a point) the positive components of S and whose lines are determined as follows: There is a line joining points π_i and π_j of the new graph is and only if there is at least one negative line in S joining a point of π_i and a point of π_j. The construction of S^* from S is illustrated in Fig. 3.2.

A signed graph S has a *unique coloring* if there is only one partition of $V(S)$ into color sets. The next theorem gives a criterion for a signed graph to have a unique coloring.

Theorem 3.2. Let S be a signed graph with a coloring. This coloring is unique if an only if S^* is complete.

The signed graph S in Fig. 3.2 is uniquely colorable. The signed graph S_1 in Fig. 3.3, on the other hand, is colorable but not uniquely, since its

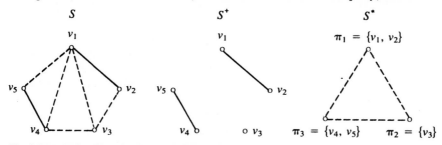

Fig. 3.2. A (uniquely) colorable signed graph.

43

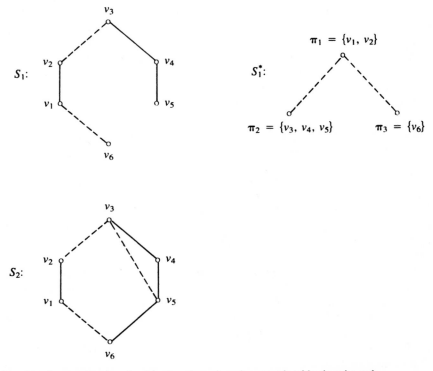

Fig. 3.3. A nonuniquely colorable signed graph and a noncolorable signed graph.

condensation S^* is not complete. Thus its point set $V(S)$ can be partitioned into the color classes $\{v_1, v_2\}$, $\{v_3, v_4, v_5\}$, $\{v_6\}$ or into the classes $\{v_1, v_2\}$, $\{v_3, v_4, v_5, v_6\}$. The signed graph S_2 in Fig. 3.3 is not even colorable.

Structural balance

In a cognitive system, if person P likes another O, P will tend to have the same attitude, positive or negative, toward some object X that O has. If P dislikes O, then P will tend to have the opposite attitude, positive or negative, toward X as O has. The same considerations would apply for positive and negative relations in general, either symmetric or asymmetric, in an interpersonal system consisting of individuals A, B, C, or in a societal system consisting of groups X, Y, Z. Situations that conform to these structures are balanced and presumably stable; those that do not are unbalanced, presumably unstable, and subject to pressure toward change. The theory of structural balance in cognitive systems was first formulated for triads by the psychologist Heider (1946), but was anticipated for interpersonal systems

Signed graphs

by Evans-Pritchard (1929), and subsequently generalized to systems with any number of elements, at any level, and formalized by Cartwright and Harary (1956). Although the theory has certain limitations in situations of competition and conflict, it does have broad application to the analysis of kinship norms and political alliance. Its formalization may be summarized as follows.

The *sign of a cycle* of a signed graph S is the product of the signs of its lines. Every positive cycle has an even number of negative lines, which of course includes the case where there are no negative lines. The signed graph is *balanced* if every cycle of S is positive. This definition agrees with Heider's classification and applies to signed graphs of any size. Fig. 3.4 shows the four 3-point signed graphs (where positive and negative values are represented by solid and broken lines, respectively), each consisting of a single cycle. The first two are balanced, the second two are not. They could, with the addition of arrows where appropriate, represent any of the situations described in the preceding paragraph.

Here are two criteria that can define balance, due to Harary (1953).

Theorem 3.3. The following conditions are equivalent for a signed graph S:

(1) S is balanced: Every cycle is positive.
(2) For each pair of points, u, v of S, all paths joining u and v have the same sign.
(3) There exists a partition of the points of S into two subsets (one of which may be empty) such that every positive line joins two points of the same subset and every negative line joins points from different subsets.

Thus a balanced signed graph is precisely one that is 2-colorable.

Returning to Brown's analysis of alliance among the Polopa in Highland New Guinea, his color metaphor corresponds to a balanced or uniquely 2-colorable signed graph. Fig. 3.5 shows a signed graph whose underlying graph is K_6, which is a subgraph of Polopa intertribal relations. As we will see later, the more general model of alliance structure for New Guinea and probably elsewhere is an n-clusterable rather than a balanced signed graph.

Structural balance occurs as a principle of kinship as well as of alliance systems. Indeed, there may even be explicit folk models of balance. Accord-

Fig. 3.4. All the signed triangles.

45

Structural models in anthropology

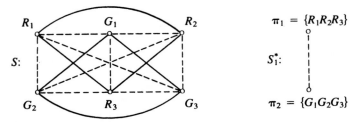

Fig. 3.5. A subgraph of Polopa intertribal alliance.

ing to G. B. Silberbauer (1972:310–11) the G/wi Bushmen of the Kalahari partition their kinship universe into two contrasting, named categories, one of which signifies permitted disrespect or familiarity, and the other enjoined respect and restraint:

Relationships between kin are categorized as either joking (*ho:khwudi*) or avoidance (*gjiukxekxu*) and an individual distinguishes between those of his kin whom he should *!ao* (fear, respect) and those with whom he may "play." In general, avoidance is expressed by: (1) Not sitting close to the relative and avoiding bodily contact; (2) Refraining from loud laughter or noisy conversation, taking care not to swear or to mention sexual or other intimate matters in the relative's hearing, and always speaking softly and politely to him or her; (3) Not addressing an avoidance relative by his or her name, but using the honorific plural, aggregate gender pronoun; (4) Not touching the avoidance relative's possessions without permission and not receiving or passing anything directly from or to the relative, but through the hand of an intermediary.

Behavior toward other kin, the joking relatives, is free of such restraint. Instead, mutual teasing, uninhibited joking and criticism, free use of one another's property, and between relatives of opposite sex and comparable age, mildly provocative flirting and a certain amount of erotic bodily contact are expected.

These relations are binary, antithetical, and symmetric. They are also exhaustive: All individuals recognized as kin are assigned to one of these two categories.

Silberbauer emphasizes the regulative and cathartic significance of this system of relations:

It contributes to the survival technique of the band by ensuring, in its operation, a free circulation of goods between joking partners and by regulating social behavior to avoid wasteful and destructive friction and tension within the kinship group. The joking relationship in allowing partners to behave towards one another in any way they please without offense being taken, provides an acceptable and harmless outlet for emotion. Inherent in the joking relationship is also a strong element of friendship which acts as a firm cohesive force within the group. The institution of avoidance, on the other hand, prescribes a type of respectful, reserved behavior which makes the development of friction between persons in this category most unlikely. (Silberbauer 1961:354)

46

Signed graphs

And he also emphasizes the fact that the G/wi Bushmen are concerned with the ways in which these dyadic relations are combined in triadic structures. Thus they say that an individual should (1) joke with the joking partner of his joking partner and (2) avoid the avoidance partner of his joking partner. But that (3) it would be bad, "embarrassing," to joke with the avoidance partner of a joking partner. If positive and negative lines of a signed graph, respectively, represent joking[2] and avoidance relations, and each point stands for a kinship status (ego and two of his kinsmen who are also kinsmen to each other), then the desirable combinations correspond to the first two (balanced) triangles in Fig. 3.4. The "embarrassing" combination corresponds to the third (unbalanced) triangle. A balanced structure consisting of more than a single triad is shown in Fig. 3.11, the G/wi atom of kinship. It is interesting to note that the balance model is doubly implied in this case, because in addition to the G/wi formulations, the ethnographer characterizes the consistency of kinship relations by the geometrical metaphor of "congruent triangles."

Not all G/wi kinship relations are balanced, however. In fact, it has been suggested (Hage 1976a) that the larger suprasystem of band alliances would be threatened were this so. There is one interesting situation in which unbalanced relations shift to balanced ones. According to Silberbauer:

There are some situations in which it [i.e., the principle of congruent triangles] does not operate, for example both cross-cousins and their parents are in the joking category, but this is changed if ego marries his cross-cousin, in which case the latter's parents are transferred to the avoidance category. (1972:311)

The situation before and after cross-cousin marriage (a stated G/wi preference) is illustrated in Fig. 3.6 (where MB = mother's brother, D = daughter, W = wife).

Finally, we observe that the balance principle extends to G/wi cosmology, as shown in Fig. 3.7, which represents a supreme being, N!adima, a lesser being, G//amama, and their relations to humans and to each other.

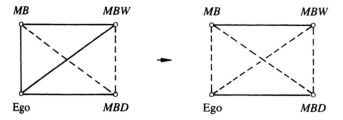

Fig. 3.6. Unbalanced and balanced structures before and after G/wi cross-cousin marriage.

[2]"Joking" and "avoidance" are Silberbauer's glosses of these G/wi relationships. These terms have multiple and variant meanings in the ethnographic literature. Our analysis does not therefore imply that all relationships which are so labeled should be treated as signed graphs.

N!adima, the creator of the universe, is "inscrutable and unapproachable by man" and is regarded with awe and fear. Humans seek only to avoid incurring his displeasure or anger. G//amama also evokes fear and sometimes cooperates with N!adima to send death and misfortune.

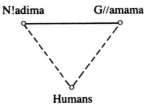

Fig. 3.7. A signed graph of G/wi cosmology.

The most famous, some would say the most infamous, treatment of kinship relations in terms of contrasting positive and negative values is Lévi-Strauss's (1963c) "atom of kinship." There are three balance analyses of this structure, one by the psychologist Claude Flament (1963) in France and two by the sociologists Peter Abell (1970) in England and Michael Carroll (1973) in Canada. Each was independently arrived at and each perhaps seemed obvious, since Lévi-Strauss used the symbols + and − to designate contrastive relations. These studies use different criteria of balance; they make different assumptions about the priority and signs of the "missing relations"; and most important, they raise the analytical and comparative problem of the coding not only of kinship relations, but of social relations in general. We emphasize that the coding problem is not restricted to social relations or graph theoretic analysis, since there has been considerable comment on the ambiguity of the + and − signs in structuralist analysis generally (see, for example, Barnes 1971).

The atom of kinship is a well-known small group consisting of the four relations brother/sister, husband/wife, father/son, mother's brother/sister's son, hereafter labeled U/M, F/M, F/S, U/S, following Flament's graph theoretic representation.[3] These express the fundamental relations of consanguinity, affinity, and filiation in a kinship system. Each of the four relations is regarded behaviorally as positive or negative. In theoretically possible systems (4 out of the combinatorially possible 16), the distribution of these signs is constrained by a rule such that there is both a positive and a negative sign both intra- and intergenerationally; that is, in each of the pairs of relations U/M, F/M, and F/S, U/S. This rule results in a transformation group, the familiar Klein 4-group (see Chapter 8) shown in Fig. 3.8, where

η = change the signs of the intragenerational lines
θ = change the signs of the intergenerational lines
ζ = change the signs of both sets of lines
ι = change the signs of neither set (which would be represented by loops in Fig. 3.8).

[3] The atom of kinship is in fact variably defined (Lévi-Strauss 1976a). We take this particular definition as convenient for the explication of balance and related graph theoretic concepts.

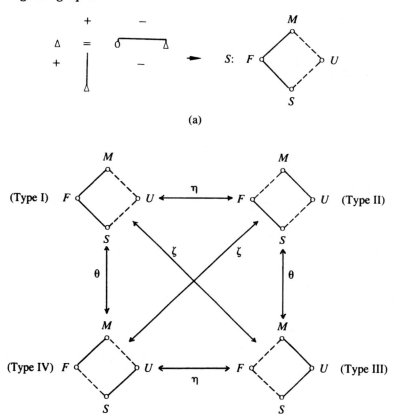

(a)

(b)

Fig. 3.8. The transformation group of the atom of kinship.

According to Lévi-Strauss, structures of types I and III are "common," whereas those of types II and IV "occur frequently but often are poorly developed." Structures with all negative signs in one of the pairs of relations (*F/S, U/S*) or (*F/M, U/M*) and all positive signs in the other are said to be "rare or perhaps impossible" because they would lead to the diachronic or synchronic breakdown of the group; that is, a disjunction between the generations or between a woman and both her husband and her brothers. (As will be evident later, the threat of breakdown in any given empirical structure would depend on how the words "positive" and "negative" are to be interpreted.)

Reasoning from balance theory, Flament observed that if the signs of two missing relations, *M/S* and *F/U*, were, respectively, positive and negative

49

and "this in all cases," then the signs of the other four relations would be the ones asserted by Lévi-Strauss, as shown in Fig. 3.9.

In the signed graphs of type α in Fig. 3.9, all four triangles, 3-cycles, are positive; in those of type β, two are positive and two are negative; in type γ, all are negative. Those in the first set are balanced; those in the second set are moderately balanced; and for those in the third set no triangle is balanced. Flament concluded:

> But essentially on the basis of observation and without any analysis referring to the theory of the balance of graphs, Lévi-Strauss...writes: Arrangements of type α are often found; on the other hand arrangements of type β are frequent but often loose, and those of type γ are rare and perhaps impossible in a clear-cut form since they would risk provoking a rupture of the elementary structure."
> It seems that the agreement between the model and reality is fairly satisfactory. (Flament 1963:126)

Two essential concepts implicit in Flament's analysis require definition: *n*-balance (Harary 1955) and the degree of balance (Cartwright and Harary 1956) of a signed graph. In many cases it may not be empirically meaningful to consider cycles of considerable length. Flament, for example, considers only 3-cycles in the signed graphs of Fig. 3.9. We therefore mention that a signed graph *S* is *n-balanced* if every cycle of length at most *n* is positive; thus Flament studies 3-balance. Obviously *S* is balanced if and only if it is *N*-balanced where *N* is the length of the longest cycle in *S*. The α signed graphs in Fig. 3.9 are 3- and 4-balanced and therefore balanced.[4] The β

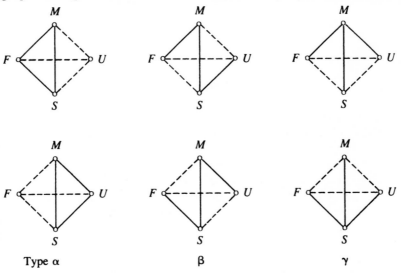

Type α   β   γ

Fig. 3.9. Signed graphs of the atom of kinship (from Flament 1963).

[4] In a complete signed graph, if all the 3-cycles are positive, all the *n*-cycles are positive.

signed graphs are neither 3- nor 4-balanced and the γ signed graphs are not 3-balanced, but every 4-cycle is positive. There are two measures of balance: a line index and a cycle index.

1. *Line index* (Harary 1959a). A signed graph S that is not balanced can be converted into a balanced one by changing the signs of some of its lines or by removing some. Intuitively, the more balanced S is, the smaller the number of lines that need to be altered. The *negation* of a sign of a line changes its sign. The next theorem provides a basis for an index of the degree of balance.

Theorem 3.4. Given a signed graph S, the minimum number of lines whose negation results in a balanced signed graph is equal to the minimum number whose deletion results in balance. Each minimal set Y of lines whose deletion results in balance can also be negated; moreover, any subset of the lines of Y can be deleted and others negated.

Based on this theorem, an *alteration minimal set* Y of lines of a signed graph S can be defined as a set whose alteration (either deletion or negation) leaves S balanced, but the alteration of any proper subset of Y does not. The *line index of balance of* S is then defined as the minimum number of lines in an alteration minimal set. The line indexes for the α, β, and γ signed graphs in Fig. 3.9 are 0, 1, and 2, respectively. There is one alteration minimal set for β which is $\{M/S\}$ and three for γ which are $\{F/M, S/U\}$, $\{M/U, F/S\}$ and $\{M/S, F/U\}$. The alteration minimal sets of S are clearly of particular interest. In the β signed graphs, for example, if as Flament remarks the sign of the line M/S is resistant to change, then other phenomena[5] might be expected, such as the preservation of an unbalanced state or the attainment of balance, but not by a principle of least effort; that is, not by altering as few lines as possible.

2. *Cycle index.* The second measure makes direct use of the definition of balance. The *cycle index of balance*[6] is the ratio β (for balance) of the number of positive cycles of a signed graph S to its total number of cycles. Thus β is the probability that a cycle chosen at random from S is positive. Obviously $\beta = 1$ if and only if S is balanced. The particular values β can take are determined by the structure of S. For the signed graphs α, β, and γ of Fig. 3.9, if only 3-cycles are considered, the values of β are 1, .5, and 0, respectively. For 3-cycles of complete 4-point signed graphs, there are no other possible values.

In the second structural balance analysis of the atom of kinship, Abell (1970:360) proposed the following interpretation of positive and negative relations and assumptions for a stable state:

[5]Flament does not specify exactly what phenomena might be expected.
[6]Although the letter β occurred in Fig. 3.9, this particular usage of it is established, and the meanings should be clear by context.

1. that the friend of my friend should be my friend;
2. that the enemy of my friend should be my enemy;
3. that the enemy of my enemy should be my friend.

He observed that if the signs of the missing relations M/S, F/U, could be predicted from the others (without, however, being regarded as necessarily derived from them), the four theoretically possible structures would be balanced in the sense of being bipartitionable. There would be positive groupings of F, M, S against U; of M, U, S against F; of F, M against U, S; and of F, S against U, M. But in no case would either of the weaker members, M or S, be a "structural isolate" against the other members of the group (see Fig. 3.10).

In the third analysis, Carroll, reasoning from balance theory, observed that the signs in the four possible structures could be only as hypothesized by Lévi-Strauss if the F/U relation were negative: The result would be the two positive 3-cycles F, M, U and F, S, U. He proposed that this relation be considered basic and the others derived. Assuming that "hostility between father and mother's brother is more likely to develop in matrilineal than in patrilineal societies," Carroll predicted that the ideal patterns should be more common in the former. Data for a cross-cultural test came from Ryder and Blackman (1970), who had previously coded the four relations in the atom of kinship as "essentially positive or negative," and for the relation F/U, from Murdock's (1967) *Ethnographic atlas* (column 22 for matrilineal descent). The prediction was not confirmed. (A second predic-

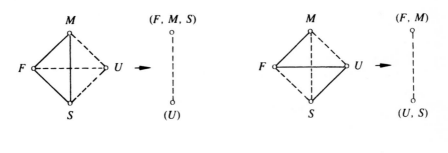

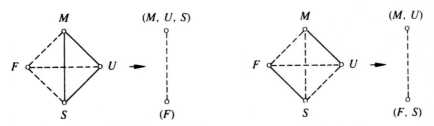

Fig. 3.10. Signed graphs of the atom of kinship according to Abell (1970).

tion, that the ideal patterns would tend *not* to occur when the *F/U* relation is positive, as inferred from the value of items exchanged in marriage, was confirmed.)

These studies call for two comments. First, they use different criteria of balance, which result in differences in the perception and analysis of structure. Cycle balance, in particular 3-cycle balance as used by Flament and Carroll, considers triads within a larger group. Partition balance, as used by Abell, considers alignments within the groups as a whole. Either criterion might be appropriate depending on circumstances, although in this particular case partition balance seems to capture the global properties with which Lévi-Strauss is concerned. We note here that the criterion of path balance (condition 2 in Theorem 3.3) has never been employed, although it promises interesting applications to the analysis of communicative consistency, that is, to the alteration and eventual sign of a message as it traverses the positive and negative lines in multiple paths joining each pair of points in a signed graph. If *S* is balanced, the message will have the same sign, whatever paths it traverses from a point *u* to a point *v*.

Second, they make different assumptions about the missing relations. There is, of course, no basis for regarding any relations as prior to or a determinant of any others. As for guessing their sign, this is dangerous. The G/wi atom of kinship shown in Fig. 3.11, for example, is balanced even though the *M/S* relation is negative, contra Flament, as is the *F/U* relation, contra Carroll.

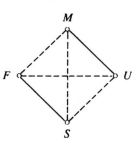

Fig. 3.11. The G/wi atom of kinship.

"Positive and negative relations"

The studies by Flament, Abell and Carroll raise a third question that is basic to any formalization of structuralism; namely, what are + and − relations? As far as the atom of kinship is concerned, it should be noted that for Lévi-Strauss, the plus and minus signs "constitute a kind of initial shorthand." For although he begins with the contrast "free and familiar relations" versus those of "hostility, antagonism or reserve," he goes on to speak of a bundle of relations comprising not only sentiment, but reciprocity and differential status as well.[7] The graph theoretic significance of this is that a system may contain heterogeneous types of relations, each

---

[7] In his later essay on the atom of kinship, Lévi-Strauss states that the signs can refer to any type of opposition: "What these attitudes are in themselves, the affective contents they mask, does not have from the particular point of view of our argument, any intrinsic meaning. In the extreme case, we would not even need to know what these contents are. It would be enough to perceive among them, directly or indirectly, a relation of opposition which the signs ' + ' and ' − ' would be enough to connote" (Lévi-Strauss 1976a:86).

of which must be coordinated to an appropriate type of graph (ordinary, signed, or directed).

Corresponding to the three types of graphs are three senses of the concept "opposite." For an undirected graph G, its *complement*, written $\bar{G}$, is obtained when the presence and absence of each line of G is interchanged. In the case of a signed graph S, its *negation, S^-,* has the same lines as S, but every sign is changed. Finally, the *converse, D',* of a digraph D is constructed by changing the direction of every arc in D. Clearly each of these three kinds of opposite is a duality; the double opposite always gives the original structure. These dualities are called, respectively, existential, antithetical, and directional. We will return to them in Chapter 6. Only the second duality is suitable for applications of balance theory. The opposite of a relation of positive sentiment is *not* a relation of superior status or an absence of reciprocity.

By way of example, Fig. 3.12 shows the Ryder and Blackman coding for Tikopia (from Firth 1957), which as in the other cases in their study is based on a mechanical or informal averaging of the heterogeneous relations of sentiment, reciprocity, and authority. The appropriate graph theoretic treatment is shown in Fig. 3.13 (where the relations F/U and M/S have been added). First, there is a signed graph S for the opposition between familiarity and restraint which the Tikopia designate as *tautau laui* versus *tautau pariki* kin relationships:

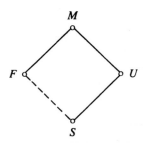

Fig. 3.12. A signed graph of Tikopia kinship relations based on data from Ryder and Blackman (1970).

The Tikopia distinguish two categories of kinsfolk of the highest importance in the regulation of the social life. These are *tautau laui* and *tautau pariki,* in literal terms, the categories of good relationship and bad relationship, but implying here not a moral judgment as to the character of the relationships themselves but a distinction

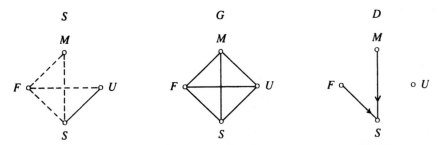

Fig. 3.13. Signed, ordinary, and directed graphs of Tikopia kinship relationships.

between the type of behaviour permissible in conducting them. Freedom in the first case, restraint in the second, are the watchwords. To the first category belong the relationships of brothers, of mother's brother and sister's child, and, to some extent, of grandparent and grandchild .... To the second category belong the relationships of parent and child, especially father and son, father's sister and her brother's children, and above all, affinal relatives. (Firth 1957:307)

Second, there is an ordinary (complete) graph G for relations of reciprocity (Firth 1957:137, 185, 198, 225, 303–4).[8] Third, there is a directed graph D for authority relations that appear to be confined to parent and child.

Three observations may be made about these graphs. (1) The signed graph in Fig. 3.12 (ignoring the added relations) is not the same as the one in Fig. 3.13. The latter is not balanced, but is clusterable (see next section), whereas the former is not. (2) Restraint and familiarity are independent of reciprocity, and negative relations do not necessarily imply authority relations. (3) The signed graph S in Fig. 3.13 represents conventional, labeled relations of familiarity and restraint. The negative signs do not imply hostility, antagonism, or enmity. There is strong positive affect between parents and children, for example, and F and U "live in amity" and F and M have "amicable" relations.

It appears, then, that care must be taken in the coordination of a graph to an empirical structure and its interpretation. In fact, one could say that graph theory, rather than leading to an oversimplification of reality, leads to a consideration of its complexity. One must understand the properties of an empirical structure before modeling it as a graph. Always, the most important features to clarify are the specification of the points (entities) and the lines (relationships).

As far as balance theory is concerned, it must be pointed out that it holds only under certain conditions. As J. A. Barnes has emphasized,[9] the basic assumption of balance theory is that solidarity is mechanical in Durkheim's sense. In kinship systems, for example, if one considers the relations between a man and his brother's wife or between a man and his wife's sister, there may be an expectation of harmony where the institutions of fraternal polyandry or sororal polygyny prevail. These institutions are rare, however, leaving plenty of room for marital discord and disharmony. In general, a significant exception to balance occurs when P and O are in competition for the same woman or the same man, or more generally for the same scarce good X.

In some circumstances, a permanent state of imbalance may occur when systems of antithetical relations are larger than a single triad. Thus if P and

---

[8]Actually, of course, reciprocity is a "polythetic" concept (Needham 1975) consisting of structures that vary according to relational properties, number, sign, and sequencing, and that may therefore require a network representation (Hage and Harary 1982).
[9]Personal communication.

Structural models in anthropology

O both have positive relations to a person R but different relations to a person S, the relation of R to S will produce an unbalanced state, whatever its sign. This is known as the "cross-pressure hypothesis," formulated by J. A. Davis (1963) and independently discovered by Evans-Pritchard. To exemplify his theory of "intersecting circles of sentiment," Evans-Pritchard describes the constellation of attitudes among parents, children, and wife's brother in certain Melanesian and African societies:

Now, in most savage societies, the wife's brother is the pivotal relative in the institutions of marriage and the family, and the attitudes of husband and wife are more pronounced towards him than towards other members of their kin and are more pronouncedly different. We may therefore be prepared to find some evidence of this clash of sentiment of the parents in the attitude of the child towards his mother's brother. Such evidence may appear in the form of strong repressed dislike of the mother's brother, such as Malinowski has unearthed in Melanesia, where it receives particular emphasis from the special institutions, modes of residence, inheritance, descent, etc., which exist in this locality. On the other hand, in tribes such as the Bathonga or the Azande, where similar institutions act far more decisively and wholeheartedly in favour of the father's line, and where consequently the friction is less severe, we find evidence for an ambivalent attitude towards the mother's brother in various forms of ritual, one of which forms is "ritual stealing." (Evans-Pritchard 1929:193)

Clustering

In his analysis of Polopa warfare, Brown implies that in the native conception, there is necessity for a structure in which the enemies of enemies (and allies of allies) are always allies, and in which there are no neutrals. As we have seen, this is realized by a signed graph that is both balanced and complete. There are, however, other groups in Highland New Guinea in which the enemy of an enemy may be either a friend or an enemy. This is an instance of what will be formally defined below as a clusterable signed graph, which is in turn a general model of alliance structure containing balance as the special case of just two coalitions. The Gahuku-Gama of the Eastern Central Highlands, as described by K. E. Read (1954), may serve as an example, an interesting one with respect to the general topic of tribal organization because in this part of New Guinea, the structure of alliance and enmity is said to define a society.

The characteristics and correlates of Gahuku-Gama warfare conform to the general picture sketched for the Central Highlands by Berndt (1964), Read (1954), and A. Strathern (1969). In form, warfare consists of both scattered raids and organized battles. It is distinguished conceptually and terminologically from fighting within the group, the latter being occasioned by specific grievances and requiring some resolution of the conflict. It is frequently a coordinated form of activity involving temporary and/or permanent alliances. It is regarded as "right, proper and commendable"; it is expected and recurrent, a normal part of social life.

As for aims and causes, warfare is sometimes precipitated by revenge, property disputes, or sorcery, but these are usually pretexts for something regarded as an end in itself. Although land pressure is the cause in more densely populated areas (Berndt 1964), the more common aim is simply dispossession of one's enemies, as well as killing as many as possible, both to demonstrate dominance and to achieve some measure of security. Since warfare is something of a "deadly game" (Berndt 1964), it is not uncommon for the victors to invite the vanquished or some segment of them to return after a period of time, with an eventual resumption of hostilities. Success in warfare is a preeminent expression of masculinity and a major means of achieving status and leadership.

With regard to frequency and effects, Berndt describes warfare in the Highlands as "chronic, incessant and endemic," although not "total." It is on a sufficient scale, however, to cause large refugee groups and considerable mobility. It is restrained to a degree by affinal links and trade relations between opposing groups. Contextually, warfare finds expression in myth, magic, and ritual, and it is related to economic and marriage exchange (in some groups warfare is identical with marriage relations and in other groups incompatible with them). It also affects socialization and settlement practices. Sociologically, in addition to generating multi-affiliation, multi-residence, and physical mobility – in short the so-called looseness of Highland social organization (Barnes 1962) – warfare is important in defining areas of intensive interaction based on common rules and conventions.

Given the ubiquity of warfare, there are a number of alternative means by which individual groups can maximize their security. These include membership in a larger group that can offer support and/or refuge, the use of neutrals to reduce the scale, multi-affiliation to gain refuge, and the formation of alliances. Among the Gahuku-Gama, the survival of a group depends on its ability to maintain an "essential balance" in its political relationships – explicitly, to avoid being surrounded by enemies but presumably also to achieve some congruence in friend-enemy relationships. Because these relationships are highly formalized and relatively permanent and because warfare is a coordinated activity, one would expect to find a high degree of consistency in their alignments.

Fig. 3.14 reproduces Read's (1954) diagram of Gahuku-Gama alliance structure. There are 16 named subtribes, two of which, Gahuku and Gama, are conjoined to designate the entire group that regards itself as a unity and that is perceived as such by adjacent groups of similar structure. The subtribes are joined by traditional relations of enmity and friendship (designated by the people as *hina* versus *rova* relations) represented by broken and solid lines, respectively.

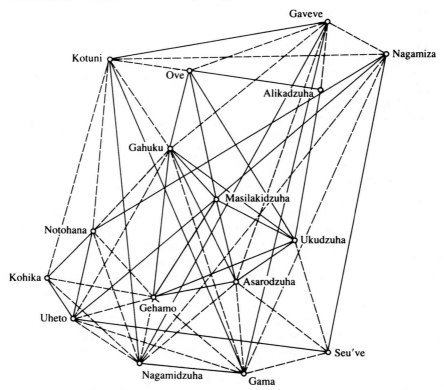

Fig. 3.14. Read's (1954) diagram of Gahuku-Gama political ties and oppositions.

Fig. 3.14 is, of course, an implicit signed graph *S*. It is immediately apparent that *S* is not balanced because it contains negative cycles, such as Notohana, Kotuni, Gahuku. It is also apparent that *S*, because it is so large, is a bit hard to interpret. Thus a better representation, in the sense of a more manipulable one, would have been as a matrix (see Chapter 5). If, however, we enumerate and classify all the 3-cycles (after adding some missing lines that emerge from Read's other ethnographic descriptions cited in Hage 1973), the results are those shown in Table 3.1.

For this signed graph, the cycle index of balance for 3-cycles is .82 (97/119). The interesting feature of Table 3.1 is that 91 percent of the imbalance is due to all-negative triangles, whereas only 9 percent is due to those with a single negative line. Empirically, the latter type of triangle is problematic: A friend of a friend is an enemy, and common friends of one party are common enemies of each other. The all-negative triangles are not: An enemy of an enemy is an enemy. We stated earlier that a balanced signed graph is one that is 2-colorable. We now define a *clusterable signed graph*

Table 3.1. *Types and numbers of each type of 3-cycle in S*

	I	II	Total
Positive	+ + + (23)	– – + (74)	97
	III	IV	
Negative	+ + – (2)	– – – (20)	22
Total	25	94	119

(J. A. Davis 1967) as one that is colorable. Clusterability is the more general model in terms of which balance is a special case. We know from Theorem 3.1 that the signed graph S of Gahuku-Gama alliance structure is not clusterable because it contains at least one cycle with a single negative line. However, S comes very close to being uniquely clusterable, as shown in Fig. 3.15 (where numbers have replaced the subtribal names).

In Fig. 3.15, every pair of points within each circle is joined by an all-positive path. The braces designate maximal complete subgraphs (cliques) within each set. With the removal of either point 7 or the lines (7,2) and (7,16), S would be uniquely 3-clusterable. The property of uniqueness immediately suggests some higher-order grouping in Gahuku-Gama society, and this is in fact the case.

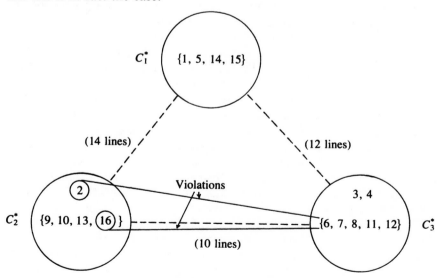

Fig. 3.15. The "almost" coloring of the signed graph in Fig. 3.12.

Structural models in anthropology

In an earlier article, Read lists "the principle, named Gahuku-Gama social groups ... and the manner in which they are combined" (Read 1951:158). At the subtribal level, 11 groups are given. Ten of these are a subset of the 16 given in the 1954 article. According to Read, the individual subtribes are merged into 3 larger opposing groups. The correspondence between these groups and the formally elucidated "almost" color classes is as follows:

Gahuku Gama groups	Almost color classes
1. Gama-Nagamidzuha (14, 15)	$C_1^* = \{1, 5, 14, 15\}$
2. Uhetove-Notohana (9, 16)	$C_2^* = \{2, 9, 10, 13, 16\}$
3. Gahuku-Gehamo (6, 7, 8, 11, 12)	$C_3^* = \{3, 4, 6, 7, 8, 11, 12\}$

Read's data, as far as they go, correspond perfectly with the "almost" color classes of S. As J. C. Mitchell has remarked in his introduction to *Numerical techniques in social anthropology,* the model enables one "not only to isolate three major alliance sets, but also to pick out those groups that are departures from the pattern and therefore present particular problems of analysis" (Mitchell 1980:16).

Three questions naturally arise in connection with balance and clustering in signed graphs: (1) When is there nonunique rather than unique clustering? (2) When is there balance rather than clustering? (3) How do balance and clustering come about? We can make a few observations here. An obvious instance of nonunique clustering occurs when one group holds a balance of power with regard to two or more other groups. An example is given by the alliance structure of Yapese society in the Western Carolines of Micronesia (Alkire 1977), as depicted in Fig. 3.16. The points 1, 2, 3, 4 represent social classes, with the even points and odd points placed on different "sides." In confrontations between the groups, the unaligned class, point 5, may side with either, depending on its calculations of expected outcomes. Similar clustering phenomena are found elsewhere in Micronesia, in the district alignments of Nauru, for example (Alkire 1977).

Two structural conditions for balance rather than clustering involve coalition formation and communicative consistency. The latter was described above in connection with the path criterion of balance. A possible environmental condition is suggested by psychological experiments that show a tendency toward polarization under conditions of stress (Taylor 1970).

There are formal models that describe balancing processes, for example Flament's (1963) lattice model of step-by-step changes from unbalanced to progressively more balanced structures. It appears, however, that these will have limited anthropological application given the kind of data usually available to fieldworkers. A more promising line of approach is simulation.

Signed graphs

Side of Young Men   Side of Chiefs

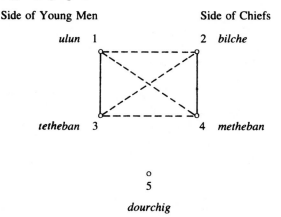

Fig. 3.16. The nonunique clustering of Yapese political alliance.

A simple but elegant example is C. R. Hallpike's (1970) analysis of the alliance structure of a set of Ethiopian towns.

Hallpike describes a system of warfare among Konso towns in which battles are fought not for ecological but for psychological reasons: Rather than being competition over scarce resources, they result from "pinpricks or minor grievances." Combatants may be helped by allies, and the distinctive characteristic of the network of friend-enemy relations is a pattern of "balanced nuclear alliances," by which is meant maximal sets of towns all of whom are friends and all of whose relations to other towns have the same sign (friendship or enmity). Most commonly these alliances contain only two members. Hallpike argues that the pattern can be explained as the outcome of random processes rather than as the result of large-town magnetism or migration. In the first case, the presumed attraction a large town has for a smaller one is contradicted by the preponderance of 2-member nuclear alliances. In the second case, the available historical evidence suggests that larger towns simply increase in size rather than fission off into new towns. To test the theory of random processes, Hallpike constructed a model network of an alliance network that was then subjected to a game sequence of battles defined by a few simple rules.

Fig. 3.17 shows the network at the start of a game. The points represent towns and the lines, relations between them. The numbers on the lines show one town's evaluation of another town, with 2, 3, and 4 signifying friendly, neutral, and unfriendly evaluations, respectively. For example, B ranks C as unfriendly and C ranks B as neutral. The numbers are assigned randomly, and there are no rankings of 1, very friendly, or 5, very unfriendly, to begin with because these are end states in the game. The two major assumptions

61

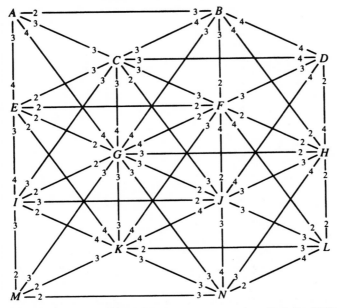

Fig. 3.17. A model network of alliance structure (from Hallpike 1970).

are that towns do not necessarily rank each other symmetrically, and that rankings change in the course of successive encounters.

The game consists of arranging battles between pairs of towns chosen at random and forcing a potential ally to choose between them. Town *A* chooses between towns *B* and *C* on the basis of its relative evaluation of each. If it is equal, then *A* chooses on the basis of *B*'s and *C*'s evaluation of *A*. If both sets of evaluations are equal, *A* remains neutral. In Fig. 3.17, in a battle between *B* and *C*, *A* chooses (allies with) *B*; in a battle between *E* and *C*, *A* chooses *C*; and in a battle between *K* and *N*, *M* remains neutral. After the battle, the participants revise their evaluations of each other: The principal opponents move down a notch toward greater enmity and so do allies on opposite sides, whereas allies on the same side move up a notch toward greater friendliness. The battles continue until all (or almost all) lines are either 1 or 5. The results of the first game played on Fig. 3.17 are shown in Fig. 3.18, and again in Fig. 3.19, in which 1 and 5 lines are replaced by solid and broken ones to clarify the pattern of friendship and enmity.

If we construct nuclear alliances using as many potential partners as possible, then in Fig. 3.19 there are three 2-member, two 3-member and zero 4-member balanced nuclear alliances: (*A, B*), (*B, F*), (*M, N*), (*E, C, G*), (*J, K, L*). After three games, the results are 9, 3, 0. In the actual Konso system the percentages of 2-, 3-, and 4-member alliances are 75, 12.5, and

Signed graphs

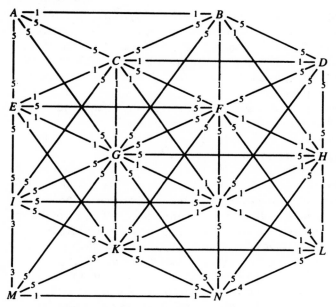

Fig. 3.18. Results of the game played on Fig. 3.17 (from Hallpike 1970).

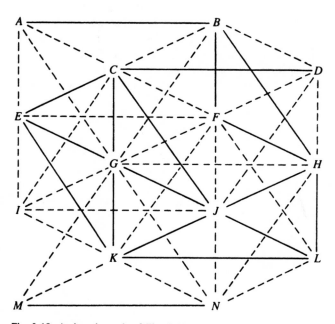

Fig. 3.19. A signed graph of Fig. 3.18.

63

12.5. The results of the simulation are close to reality. The more frequent occurrence of 2- as opposed to 3- and 4-member alliances is explicable on the basis of less likelihood of contradictory relations between alliance members. Hallpike also finds that the simulation is comparable to the real world because both systems contain unbalanced relations, isolates, and chains of nonnuclear alliances.

Related statistical formulas have been developed by Frank and Harary (1980) to measure the amount of balance in a randomly generated signed graph. Both presence versus absence of lines and the signs ($+$ or $-$) of those lines that are present have a given probability, say 0.5.

The model of balanced signed graphs has also found application in subjects outside anthropology and cognitive social psychology. Studies of anthropological interest include Roberts's (1976, 1979) work on the instability of causation structures defined by economic and environmental variables, Harary's (1961a) analysis of shifting alignments in the Middle East, Healy and Stein's (1973) favorable evaluation of balance theory as compared to a variety of balance of power theories with reference to the Bismarckian epoch of international relations, and Cartwright and Harary's (1959) analysis of the attainment of equilibrium in Freud's theory of ego development, their overview (1979) of balance theory, and Harary's (1983) appraisal of consistency theory.

Signed graphs provide the appropriate models for the analysis of antithetical relations. The theories of structural balance and clustering principles, which are based on the coloring of signed graphs, serve to characterize patterns of consistency in alliance and kinship systems. Because not all relations, signed or unsigned, are symmetric, but may be either asymmetric or nonsymmetric, we next introduce directed graphs as structural models.

4

Digraphs

My problem is simple. How can a modern social anthropologist, with all the work of Malinowski and Radcliffe-Brown and their successors at his elbow, embark upon generalization with any hope of arriving at a satisfying conclusion? My answer is quite simple too; it is this: *By thinking of the organizational ideas that are present in any society as constituting a mathematical pattern.*

Edmund Leach, "Rethinking anthropology"

In a modern classic of New Guinea ethnography, Raymond Kelly (1974) develops for the analysis of Etoro social structure a logical approach that is clearly meant to have general application. The logic is based on an analysis of the concept of siblingship, which he defines by a principle of transitivity or transitive equivalence:

The recognition of siblingship as a principle of relationship (rather than merely a unitary element of structure) necessitates a redefinition of the term, and one that will effectively extract siblingship from the familial context which has heretofore served as its primary referent. This can be accomplished by drawing on the relational concept of transitivity (Feibleman and Friend 1945:21). A relationship between two elements (X and Y) is transitive when it is contingent upon, and in this sense defined by, their respective relations to a third element (M). (Transitive relations are thus inherently indirect, being mediated by a third element.) I will be concerned with only one specific type of transitivity, that in which the relationship of each of the two elements (X and Y) to the mediator is identical. To be perfectly explicit, I am concerned with situations in which (1) the relation between X and Y is defined by the relation of X to M and of Y to M, respectively, and (2) X and Y are each related to M in the same way (i.e., X:M::Y:M). (Kelly 1974:268)

According to Kelly, this principle is implicit in Meyer Fortes's well-known characterization of siblingship as a relation that "connotes common parentage, which in terms of the calculus of filiation means co-filiation" (Fortes 1969:270).

To refer siblingship to "common parentage" is to say that the relation between any two siblings (X and Y) is mediated by their respective (equivalent) relations to a third element, the parental pair (M), i.e., to invoke a transitive relation. (Kelly 1974:269)

But Kelly also regards this principle as a generalized, abstract one. "Transitive equivalence" among siblings can thus be mediated affinally as well as consanguineally, as in fraternal polyandry and sororal polygyny, when same-sex siblings have identical relations to a common spouse and to their spouse's kin. His principle covers "non-genealogical brotherhood" and even more generally it defines "congruence in individual kin networks" by which is meant "the condition whereby two (or more) individuals have relatives in common, although both individuals need not be related to these relatives in precisely the same way" (Kelly 1974:274). Transitive equivalence can be mediated by marriage rules – exogamy and antigamy – resulting in "equivalent relationships to an external source of women." And it can be mediated materially by co-ownership of territory.[1] Most significantly for New Guinea studies, the principle of transitive equivalence, which is also called the principle of siblingship, is applicable to notions of descent, an immediate consequence of which is a dissolution of the problem of "loose structure":

It may be pointed out that siblingship subsumes certain aspects of descent; utilization of the concept therefore has the potential for resolving a discrepancy between ideology and statistical norms whenever such discrepancy arises from identification of an ideology as "patrilineal" on the grounds that co-members of groups and/or aligned segments are said to be "brothers" although statistically they are not genealogically related as patrilineal descendants of the same ancestor (or hierarchy of ancestors). In other words, co-membership and alignment phrased in terms of "ideological brotherhood" is not necessarily inconsistent with omnivorous recruitment and the resultant presence of significant numbers of genealogical nonagnates in local groups. On the contrary the principle of siblingship may uniformly order both relations between members of local groups and the alignment of territorial segments. Indeed, it appears to be a general feature of "loosely structured" New Guinea societies that "brother" relationships are preeminently mediated by territory (irrespective of ancestry) and the fact that brotherhood is imputed between agnates and genealogical nonagnates who are (or who have come to be) co-owners of the territory of the local groups is indicative of this. (Kelly 1974:277–8)

The shortcoming of this analysis is that Kelly's conception of transitivity does not coincide with the accepted use of that term in logic. Nevertheless, the type of model he seeks to develop is of considerable importance for Melanesian and for social structure studies in general. It is an application of the formal analysis envisaged some time ago by Lévi-Strauss. As he observed, "with the help of such notions as transitivity, order, and cycle, which admit of mathematical treatment, it becomes possible to study, on a

---

[1] One might expand this list structurally by including the ascending relation of teknonymy, in which parents derive a common name from their child. In Balinese society, for example, this practice identifies the husband-wife pair, reflects the symmetry of this relation, and underlines its importance as the "fundamental social building block" (Geertz 1973).

66

purely formal level, generalized types of social structure, where both the communication and subordination aspects are fully integrated'' (Lévi-Strauss (1963a:312). These notions are fundamentally graph theoretic ones.

The theory of relations has very general significance in anthropology. In addition to the formal analysis of social structure, logical relations serve to discriminate basic types of marriage exchange (Lévi-Strauss 1969), define consistency in kinship rank (Hage 1976b), establish universals in language (Greenberg 1963, 1978), in the sexual division of labor (Burton, Brudner, and White 1977), and in primitive classification (Berlin and Kay 1969; Berlin, Breedlove, and Raven 1973), distinguish types of evolutionary sequences (Carneiro 1973), articulate categories of belief systems (D'Andrade 1976), order ritual structures (Atkins and Curtis 1969), and characterize basic cognitive operations (Wallace 1962).

Logical relations, however, are not always fully explicit, or consistently applied in anthropology. For example, in Brown's models of New Guinea social structure, cited at the beginning of the previous chapter, his use of transitivity implies different logical concepts from Kelly's use here, and neither author intends what Wilson (1980) does in his transitivity model of the evolution of human social structure. There are also problems arising from the effect of number on the structure of a relation. One of the advantages of graphical representation is that it provides clear and precise definitions of a large repertory of relational concepts.

Directed graphs or digraphs are graphs whose pairs of points are ordered or graphs whose lines have arrows. Digraphs allow for asymmetric as well as symmetric relations and are therefore more complex and in some respects more interesting than graphs. Our presentation begins with a set of definitions parallel to those given for graphs in Chapter 2. Then, using graph theory as a model, we define seven basic properties of relations – reflexivity, irreflexivity, symmetry, asymmetry, transitivity, intransitivity, and completeness – and use these properties to axiomatize the several kinds of relations, including digraphs, graphs, equivalence relations, parity relations, partial orders, and complete orders. The clarification of Kelly's concept of transitivity shows it to be a parity relation. We then introduce acyclic digraphs and, using the concepts of transitivity and order, provide a model for the characterization of consistency and hierarchy in kinship relations. The ethnographic examples are from Polynesian and Micronesian chiefdoms. A digraph of a complex structure can be simplified by shrinking it, technically by condensing it, as exemplified by stratification in a Melanesian men's society. This leads naturally to a comparison of Melanesian and Micronesian political hierarchies interpreted as semilattices and trees. Finally, we illustrate the operation of cyclic relations and theories of indirect causality in classification systems by means of the Five-Element Theory in Chinese philosophy.

Structural models in anthropology

Definitions

A *directed graph* or *digraph D* consists of a finite set *V* of points and a collection of ordered pairs of distinct points. Any such pair (*u*, *v*) is called an *arc* or *directed line* and will usually be denoted *uv*. The arc *uv* goes from *u* to *v* and is incident with *u* and *v*. We also say that *u* is *adjacent to v* and *v* is *adjacent from u*. In a graph *G* each point has a degree, but in a digraph *D* it has both an outdegree and an indegree. The *outdegree od* (*v*) of a point *v* is the number of points adjacent from it and the *indegree id* (*v*) is the number of points adjacent to it. Fig. 4.1 shows all the nonisomorphic digraphs with three points. In the penultimate digraph of the first row, there is one point where *od* = 1 and *id* = 2, one where *od* = 2 and *id* = 2, and one where *od* = 2 and *id* = 1.

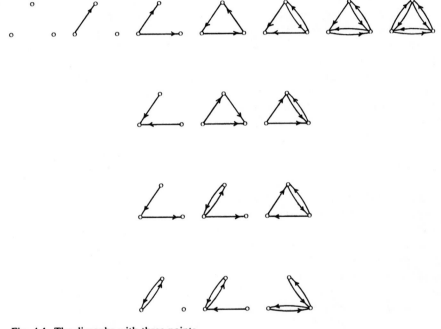

Fig. 4.1. The digraphs with three points.

A (*directed*) *walk*[2] in a digraph is an alternating sequence of points and arcs, $v_0, x_1, v_1, \ldots, x_n, v_n$ in which each arc x_i is $v_{i-1}v_i$. For brevity, we may write the sequence of points $v_0, v_1, \ldots, v_n$ to indicate the same walk. The *length* of such a walk is *n*, the number of occurrences of arcs in it. A *closed walk* has the same first and last points, an *open walk* does not, and a

[2] Unfortunately, a walk was called a "sequence" when it was defined in Harary et al. (1965:39) and throughout that book. An arc was also called a "line" there.

68

Digraphs

spanning walk contains all the points. A *path* is a walk in which all points are distinct; a *cycle* is a nontrivial closed walk with all points distinct except the first and last. If there is a path from u to v, then v is said to be *reachable from u*, and the *distance d* (u, v) from u to v is the length of any shortest such path. In Fig. 4.2, the sequence of points 1, 2, 3, 4, 2, 5 is a directed walk; the sequence 1, 2, 5 is a path; and 2, 3, 4, 2 is a cycle. There is no spanning walk.

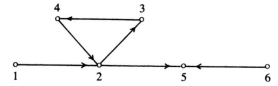

Fig. 4.2. A digraph to illustrate walks.

Each walk is directed from the first point v_0 to the last v_n. We also need a concept that does not have the property of direction and is analogous to a walk in a graph. A *semiwalk* is again an alternating sequence $v_0, x_1, v_1, \ldots, x_n, v_n$ of points and arcs, but each arc x_i may be either $v_{i-1}v_i$ or v_iv_{i-1}. A *semipath* is a semiwalk in which all points are distinct, and a path is a semipath with consistent direction. A semipath that is not a path is called a *strict semipath*. In Fig. 4.2, points 1 and 6 are *joined* by a semiwalk (1, 2, 3, 4, 2, 5, 6) and by a semipath (1, 2, 5, 6). A *semicycle* is obtained from a semipath on adding an arc joining the terminal and initial points of the semipath. Every cycle is a semicycle, since every path is a semipath, but not conversely.

Whereas a graph G is either connected or it is not, there are three different ways in which a digraph may be connected. A digraph is *strongly connected*, or *strong*, if every two points are mutually reachable; it is *unilaterally connected*, or *unilateral*, if for any two points, at least one is reachable from the other; and it is *weakly connected*, or *weak*, if every two points are joined by a semipath. Clearly, every strong digraph is unilateral and every unilateral digraph is weak, but the converse statements are not true. A digraph is *disconnected* if it is not even weak. We note that the trivial digraph consisting of exactly one point is (vacuously) strong because it does not contain two distinct points. There are four different connectedness categories of a digraph D which indicate its highest level of connectedness. Fig. 4.3 shows digraphs that are *weak but not unilateral* (C_1), *unilateral but not strong* (C_2), and *strong* (C_3). Each digraph is intentionally drawn with the same number of arcs (and points) to show the lack of a direct relation between structure and density, as emphasized for graphs in Chapter 2. Finally, C_0 is the category of disconnected digraphs, those that are not even weak.

69

Structural models in anthropology

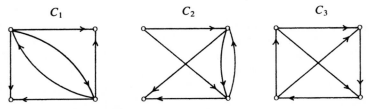

Fig. 4.3. Digraphs to illustrate connectedness category.

In contrast to the connected components of a graph, there are three different kinds of digraph components: strong, unilateral, and weak subgraphs that are maximal with respect to these properties.

Relations

A rigorous definition of a relation requires some basic concepts from set theory. A *set S* is just a collection of elements. For example, the alphabet A is the set of letters a, b, ..., z. It is customary mathematical notation to write $A = \{a, b, c, \ldots, x, y, z\}$, which is read "A is the set consisting of the elements $a, b, c, \ldots$." An *ordered pair*, written (x, y), is the set of two elements with the additional proviso that x is called the first element and y the second. In other words it is a 2-term sequence. An *ordered triple* (x, y, z) is, as expected, a 3-term sequence. A *relation R on a set V* is precisely a set of ordered pairs of elements of V. The *domain* of a relation R is the set consisting of all the first elements of the ordered pairs in R; its *range* is the set of all second elements.

The historical origins of this formulation go back to the philosopher-logician Charles Peirce (1931), who interestingly enough illustrated these concepts using kinship relations. Thus in English, the relation "father" is the set of all ordered pairs (x, y) such that x is the father of y; its domain is the set of all men who have children; and its range is the set of all children. "Brother" is a relation whose domain is the set of all males who have a sibling, and its range is the set of all people who have a male sibling. "Uncle" is a composite relation, being the union of the two relations, "brother of a parent of" and "brother-in-law of a parent of." Each of these two is called a *relative product*.

A relation R is represented graphically by taking the elements of V as points and drawing an arc from u to v whenever the ordered pair (u, v) is in R, and also an arc from point u to itself, called a *loop*, if (u, u) is in R. Fig. 4.4 shows a relation R

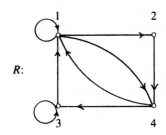

Fig. 4.4. A small relation.

70

on $V = \{1, 2, 3, 4\}$ having eight arcs, two of which are loops; here $R = \{(1, 1), (1, 2), (1, 4), (2, 4), (3, 1), (3, 3), (4, 1), (4, 3)\}$.

Because a relation is a set, and as no set lists any of its elements more than once, there can never be more than one arc from one point to another (although there may be two arcs, one in each direction, as in Fig. 4.4). It is most convenient and intuitively clear to exploit the graphical representation of a relation when developing the rich variety of properties of relations. Fig. 4.5 shows some of the relations on three points.

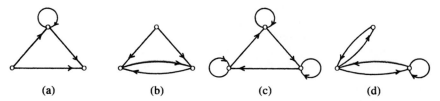

(a)　　　　　　(b)　　　　　　(c)　　　　　　(d)

Fig. 4.5. Relations on three points.

Relation R is *reflexive* if there is a loop at every point; it is *irreflexive* if no point has a loop. In Fig. 4.5, (c) is reflexive and (b) is irreflexive; (a) and (d) are neither. As an empirical example, if we represent two or more social groups as points and the relation "intermarries" as lines, endogamy determines a reflexive relation and exogamy an irreflexive one.

Whenever there are two arcs of the form (u, v), (v, u), they form a *symmetric pair* of arcs. Then R is symmetric if every arc is part of a symmetric pair; R is *asymmetric* if no arc is part of a symmetric pair. In Fig. 4.5, (d) is symmetric and (a) and (c) are asymmetric; (b) is neither symmetric nor asymmetric. In Lévi-Strauss's (1969) classification of marriage systems, restricted exchange is a symmetric relation and generalized exchange is an asymmetric relation.

The next property of a relation depends on the concept of a *2-path* from u to v. This consists of two arcs of the form (u, w) (w, v) with the three points u, v, w distinct. Then relation R is *transitive* if, whenever there is a 2-path from u to v, the 1-path or arc (u, v) is also in R. By an *intransitive* relation is meant one that is never transitive; that is, whenever there is a 2-path from u to v, the arc uv cannot be present. In Fig. 4.5, (a) and (b) are transitive; (c) and (d) are intransitive. An example of a transitive relation occurs in syllogistic reasoning, a universal cognitive operation (Wallace 1962) based on implication. If a implies b and b implies c, then a implies c; symbolically $a \rightarrow b$ and $b \rightarrow c$ imply $a \rightarrow c$.

Perhaps the most straightforward application of transitivity is to structures of subordination. A contrast has been proposed, for example, between the transitive aspects of the dominance structure of hens, the pecking order, and the transitive structure of human societies:

71

Structural models in anthropology

Of a very promising nature for study of social structure are Rapoport's attempts to formulate a mathematical theory of the pecking order among hens. It is true that there seems to be a complete opposition between, let us say, the pecking order of hens, which is intransitive and cyclical and the social order (for instance the circle of *kava* in Polynesia), which is transitive and non-cyclical (since those who are seated at the far end can never sit at the top). But the study of kinship systems shows precisely that, under given circumstances, a transitive and non-cyclical order can result in an intransitive and cyclical one. This happens, for instance in a hypergamous society, where a circular marriage system with mother's brother's daughter leaves at one end a girl unable to find a husband (since her status is the highest) and at the other end a boy without a wife (since no girl, except his sister, has a status lower than his own). Therefore either the society under consideration will succumb to its contradictions, or its transitive and non-cyclical order will be transformed into an intransitive and cyclical one, temporarily or locally. (Lévi-Strauss 1963a:311)

In fact, Lévi-Strauss's characterization of the pecking order is logically correct only when the flock contains exactly three hens, u, v, w, which peck each other cyclically rather than transitively. The moment the flock has four or more hens, the resulting pecking order digraph D cannot logically be intransitive, as the following argument for four hens, u, v, w, x, demonstrates. Without loss of generality, begin by assuming the first three hens peck each other in the cyclic order $u\ v\ w\ u$, so that in particular the arc (u, v) is in D. Now we try to make all four triples cyclic and see that this cannot be done. Consider next the three hens u, v, x. As it is already known that u pecks v, the cyclic ordering is fully determined, $u\ v\ x\ u$. Then, as Fig. 4.6 shows, no matter which of w and x pecks the other, the remaining two triples in D must be transitive. Hence D cannot be intransitive, but is not transitive.

Actually this observation is a special case of a general formula due to Kendall and Smith (1940), which proves at once that no flock with $p \geq 4$ hens can be intransitive, as every such flock must have a certain number of transitive triples; see

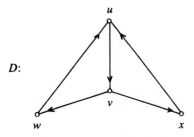

D:

Fig. 4.6. Impossibility of intransitivity.

Harary 1969:207, Harary et al. 1965:300. Thus, by intransitive Lévi-Strauss really means not transitive, but in fact this is not an accurate characterization of the pecking order. According to Landau (1951), the pecking order is usually transitive but not always. Occasionally, a flock may contain a cyclic triple (u, v), (v, w), (w, u), but this is a rare phenomenon in pecking orders of hens; most pecking orders are transitive, and all are complete and asymmetric.

Another useful property of relations is that of completeness. A relation R

is *complete* if for each pair of points u, v of R, at least one of the two arcs (u, v) and (v, u) must occur. The best-known empirical realization is that of the pecking order of a flock of hens in which for every pair of hens, exactly one of the two arcs appears in R.

In ethnosemantic studies, the usual assumption has been that categories are hierarchically ordered by class inclusion, a transitive relation. For example, a black oak is a kind of oak, an oak is a kind of tree, and a black oak is a kind of tree. There do, however, appear to be significant exceptions to this rule, intransitive relations underlying assertions such as these: A scrub oak is a kind of oak, an oak is a kind of tree, but a scrub oak is *not* a kind of tree but a kind of bush (Kay 1975; Randall 1976). Similarly, in psychological studies individuals sometimes display intransitive preference hierarchies such that they prefer x to y and y to z, but z to x and not x to z. The study of intransitive relations poses important problems for theories of semantic memory, one of whose aims is to account for logical relations obtaining among concepts (Hampton 1982), and for theories of choice based on notions of rational behavior (Tversky 1969).

There are potentially interesting and significant contrasts between the logical properties of actual and mythical structures within the same society, as Lévi-Strauss has pointed out:

It is also possible to enlarge the field of enquiry and to integrate for a given society, actual and potential types of order. For instance in human societies, the actual forms of social order are practically always of a transitive and non-cyclical type: If A is above B and B above C, then A is above C; and C cannot be above A. But most of the human "potential" or "ideological" forms of social order, as illustrated in politics, myth, and religion, are conceived as intransitive and cyclical; for instance in tales about kings marrying lasses and in Stendahl's indictment of American democracy as a system where a gentleman takes his orders from his grocer. (Lévi-Strauss 1963a:312)

It appears that intransitive cyclical structures can provide real as well as imaginary solutions to problems of status asymmetry, as illustrated by marriage among the Canelos Quichua of eastern Ecuador. According to Whitten (1976), a man is "subordinate" or "indebted" to his wife's brother, as shown in Fig. 4.7(a). Two ways out of this evidently unpleasant state of affairs are sister exchange [Fig. 4.7(b)] and the marriage of one's sister to the superordinate of a superordinate [Fig. 4.7(c)], thus producing a "series of equalizing asymmetries" (Whitten 1976:114).

Some relations may be more transitive than others. For example, the relation in Fig. 4.8(a) has two transitive triads, whereas the relation in Fig. 4.8(b) has three. Various measures of transitivity are given in Harary and Kommel (1979), and there are statistical formulas for transitivity that enable one to compare sets of social or cultural structures.

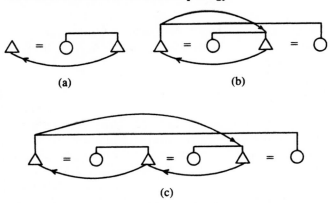

(a)   (b)

(c)

Fig. 4.7. Solutions to status asymmetry in Quichua marriage.

Axiomatization of relations

A digraph is just an irreflexive relation; there are no (directed) loops. A graph is an ir-reflexive symmetric relation. (These two sentences can serve as the basis for axiom systems for graphs and digraphs.) It is customary to draw a sym-metric relation by replacing each symmetric pair of arcs by an undirected line, as in Fig. 4.9, and as we have done throughout Chapters 2 and 3.

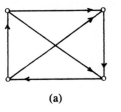 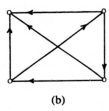

(a)   (b)

Fig. 4.8. More and less transitive relations.

u   *v*   *u*   *v*

(a)   (b)

Fig. 4.9. A symmetric pair of arcs (a) and a line (b).

It is convenient to use the following abbreviations for the seven properties of relations first introduced:

r	reflexive	*t*	transitive
r̄	irreflexive	*ī*	intransitive
s	symmetric	*c*	complete
s̄	asymmetric		

Several kinds of relations can be axiomatized using these seven proper-ties. Two have just been mentioned: A digraph is a relation that satisfies just axiom *r̄*; for a graph, axioms *r̄* and *s* hold (Table 4.1). To add one

Table 4.1. *Axiom systems for relations*

Kind of relation	Axiom system
Digraph	$\bar{r}$
Graph	$\bar{r}, s$
Equivalence relation	r, s, t
Parity relation	$\bar{r}, s, t$
Partial order	$r, \bar{s}, t$
Complete order	$\bar{r}, \bar{s}, t, c$

axiom at a time, a *parity relation* (Harary 1961c) satisfies the three axioms, $\bar{r}$, s, and t, so that it is a transitive graph. A parity relation is a graph in which every connected component is complete. The classical concept of an *equivalence relation* (r, s, t) is obtained from a parity relation by adding a directed loop at each point (the reverse process in fact suggested the concept of a parity relation). The three axioms (which fortuitously occur in alphabetical order) r, s, and t apparently first appeared as axioms for plane geometry in Euclid's classic work, in connection with the lengths of line segments.

A *partial order* ($\bar{r}$, $\bar{s}$, t) differs from a parity relation in one axiom; s is replaced by $\bar{s}$. A partial order is often called a *partially ordered set* or *poset*. The typical partial order in mathematics is that of a *proper subset*, which we denote here for two sets A and B by $A < B$; that is, A is contained in B but is not identical with B. We verify that this relation is a partial order:

$\bar{r}$ $A < A$ never holds as $A = A$
$\bar{s}$ if $A < B$, then $B < A$ is impossible
t if $A < B$ and $B < C$, then of course $A < C$.

This is called a partial order because there may well exist two sets A and B for which neither $A < B$ nor $B < A$ hold. This deficiency is removed by introducing a *complete order* ($\bar{r}$, $\bar{s}$, t, c), which is simply a partial order that is also complete.

These properties can be illustrated using kinship relations. The relation "spouse" is $\bar{r}$, as nobody is married to him or herself, not even in San Francisco. It is also s, since if u is the spouse of v, then necessarily v is the spouse of u. Funnily enough, this relation is also t (in a monogamous society but not in a polygamous one). This relation is called *vacuously transitive* because the very definition of transitivity is that the existence of a 2-path from u to v implies the existence of the 1-path as well, and there are, of course, no 2-paths in R!

Consider for illustrative purposes a family of five consisting of a husband, h, and wife, w, having two female children f and f' and one male child m. Fig. 4.10(a) shows the relation "spouse" in this family, 4.10(b)

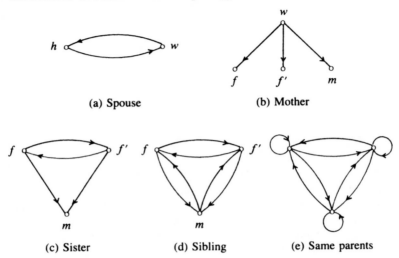

(a) Spouse   (b) Mother

(c) Sister   (d) Sibling   (e) Same parents

Fig. 4.10. Same family relations.

that of "mother," 4.10(c) that of "sister," 4.10(d) "sibling," and 4.10(e) "has the same parents." A few comments about Fig. 4.10 will clarify some concepts. The first four diagrams are digraphs; the fifth, since it has loops, is a relation that is not a digraph. "Mother" is $\bar{r}$ and $\bar{s}$, "sister" is $\bar{r}$ and t but neither s nor $\bar{s}$, "sibling" is $\bar{r} \, s \, t$ and so is a parity relation; "same parents," being $r \, s \, t$, is an equivalence relation.

We return now to Kelly's principle of transitive equivalence. He considers that this principle is a generative, pervasive model, similar to the type evoked by Lévi-Strauss (1960) in his answer to an empiricist critique of structuralism. In Lévi-Strauss's famous metaphor, the analytical aim is not to find out how the pieces of a jigsaw puzzle fit together, but rather to find the "mathematical formula expressing the shape of the cams and their speed of rotation," which determines the shape of the pieces. This is a deeply intriguing metaphor, any serious consideration of which would require some very explicit mathematics.

Kelly derives his definition from Feibleman and Friend (1945:21), who say: "transitivity is such that if it relates two parts to a middle part, it relates the extreme parts to each other." In this connection we note that one of the axioms of Euclid's *Elements of geometry* is remarkably similar: "Things equal to the same thing are equal to each other." In fact, these two quotations become identical under the correspondence:

        parts – things
        relate – equal
  middle part – same thing

76

Digraphs

To be precise, this is not the meaning of transitivity, which is correctly given by the earlier definition, "relation R is transitive if whenever there is a 2-path from u to v, the 1-path or arc (uv) is also in R." Kelly's definition that a "relationship between two elements (X and Y) is transitive when it is contingent upon, and in this sense defined by their respective relations to a third element (M)" is plainly incorrect. This is so because only a binary relation, which by definition is a set of ordered pairs, and not just a single ordered pair, can be transitive and not a relationship between two elements. If this had been stated correctly, it would have been just another rewording of the quotes above.

Now what might Kelly really mean? It appears that what he is trying to get at is the concept of a parity relation defined above (Harary 1961c), in the following way. When he says, "To be perfectly explicit, I am concerned with situations in which (1) the relation between X and Y is defined by the relations of X to M and of Y to M, respectively, and (2) X and Y are each related to M in the same way (i.e., X:M::Y:M)," it is clear that M is not playing the role of a point between X and Y in a directed 2-path. Rather, M has the role of a kind of binding rule. This means that any two elements in a relation to M in this undefined way are meant to be the points of a symmetric pair of arcs. If, following Fig. 4.9, each symmetric pair is replaced by an undirected line, then three elements X, Y, Z, all of which are related to M, would form a triangle in which M itself does not appear. And if there are four elements W, X, Y, Z, all related to M, then we would have the complete graph K_4 shown in Fig. 4.11(b). What M really does is to always give a parity relation and hence one that is both symmetric and transitive. As mentioned above, every parity relation is realized by an undirected graph in which each connected component is complete. The relation of sibling-ship is precisely a parity rela-tion.

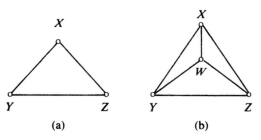

(a)   (b)

Fig. 4.11. Graphs of parity relations with three and four elements.

Two related operations concerned with transitivity in digraphs should be defined here: the transitive closure and the cover of D. The second opera-tion provides a means for simplifying a transitive digraph, and it also illus-trates the linearity of an asymmetric transitive relation.

A digraph that is not transitive can be made transitive by forming its *transitive closure*, D^t, defined as the minimum transitive digraph containing D and having the same set of points as D. To construct D^t, we add to D all arcs not already in D whenever there is a path of length 2 in D from u to v

and iterate this procedure until no new arcs can be added, as illustrated in Fig. 4.12. Equivalently, D^t is obtained from D in one step on adding arc (u, v) whenever v is reachable from u.

A transitive digraph D of a complex empirical structure can be simplified by finding the smallest possible subgraph E of D whose transitive closure E^t is all of D. We now describe the construction of E from D and call E the *blanket* of D. We say that one point u *covers* another point v if arc (u, v) is in D, but there is no 2-path from u to v. If D is transitive, then the transitive closure of the blanket of D is D itself. This is illustrated in Fig. 4.13, which shows a digraph that is t and also $\bar{s}$ and c. In such a case, the blanket is a directed path. This illustrates what was meant in Chapter 2 when such struc-

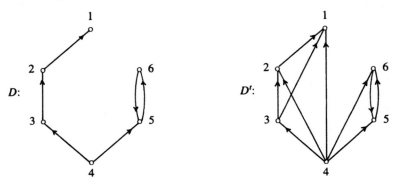

Fig. 4.12. A digraph D and its transitive closure D^t.

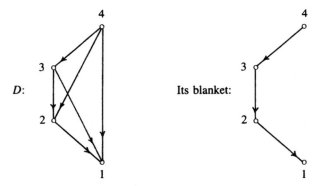

Fig. 4.13. A transitive digraph and its blanket.

tures were said to be linear. One can think of the blanket as the result of eliminating all the transitively derivable arcs. This method of simplification has been used in the analysis of ritual and belief systems (Atkins and Curtis 1969; D'Andrade 1976) and could be used for kinship digraphs like those shown in Fig. 4.16.

Digraphs

Consider the set $N_4 = \{1, 2, 3, 4\}$ of the first four positive integers as V, and the relation "greater than," written $a > b$, as relation R. This is a complete order, as it is $\bar{r}\ \bar{s}\ t\ c$. Its diagram, which is also Fig. 4.13, is a special case of a tournament.

For two sets S and S', we say S is a *subset* of S' and write $S \subset S'$ if every element of S is in S'. Hence every set is a subset of itself. Now consider two elements, a, b, and all four subsets of $S = \{a, b\}$, including S itself, $A = \{a\}$, and $B = \{b\}$, each called a *singleton*, as it is a 1-element set, and the empty set containing no elements, conventionally written $\emptyset$. Let V be the set $\{S, A, B, \emptyset\}$ whose elements are themselves sets, and let R be the relation "subset

defined on V," as shown in Fig. 4.14, in which the custom of lattice theory is invoked that whenever a line appears, its lower point is a subset of its higher point.

A partial order was defined as an $\bar{r}\ \bar{s}\ t$ relation, but it has also sometimes been defined by $r\ \bar{s}\ t$. It would be better and clearer if partial order were always used to mean an $\bar{r}\ \bar{s}\ t$ relation. Then the $r\ \bar{s}\ t$ case should be called a *reflexive partial order*. The transitive closure of Fig. 4.14 gives a partial order. It is not a complete order, since neither set A nor B contains the other.

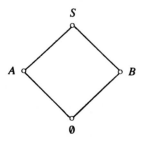

Fig. 4.14. The "lattice" of subsets of the 2-element set, $S = \{A, B\}$.

A *tournament* is a complete, asymmetric digraph; thus it is a relation satisfying $\bar{r}\ \bar{s}\ c$. Since a partial order is $\bar{r}\ \bar{s}\ t$, a complete order, $\bar{r}\ \bar{s}\ t\ c$, is a partial order that is complete; it has no pair of incomparable elements. On the other hand, a *transitive tournament* is $\bar{r}\ \bar{s}\ t\ c$, so it is identical with a complete order. All the tournaments with three and four points are shown in Fig. 4.15. The terminology arose from "round robin tournaments" in which each pair of players (or teams) in V play just once, and a draw is not

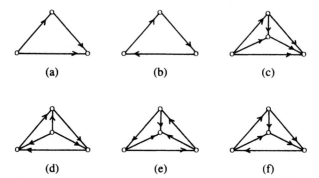

Fig. 4.15. Some small tournaments.

79

permitted. In Fig. 4.15, (a) is called the *transitive triple* and (b) the *cyclic triple*. In Fig. 4.15, (c) is just an isomorphic presentation of *D* in Fig. 4.13, and (d) and (e) are converses of each other.

Tournaments are a pervasive mathmetical model, occurring in studies of consumer preferences and other preference relations, precedence relations, scheduling problems, and paired comparisons generally. Surveys of the field include Moon (1963) and Harary et al. (1965). Tournaments also occur in anthropology in structures of subordination, as illustrated in the next section.

Acyclic digraphs

We now use the preceding definitions of relations, orders, and tournaments, together with a new concept, level assignment, in acyclic digraphs to analyze status in kinship and political systems. We also present another method of structural simplification: Just as a signed graph can sometimes be reduced to its clusters, so a digraph can be condensed into its strong components. The ethnographic examples are kinship structures from Polynesia and Micronesia and a political system from Melanesia. The latter serves in addition to illustrate the applicability of graph theoretic models to informally as well as formally defined social relations.

Irving Goldman, in his comparative study of Polynesian social structure, makes explicit reference to Lévi-Strauss's essay on the atom of kinship and suggests that for Tonga (and for other Polynesian societies), the essential feature of the relations is asymmetry based on the honor that ego owes to alter: "These kinship patterns are, in fact, organized around honour rather than around the more general sentiments of freedom and restraint" (Goldman 1970:453). This suggests that the principle of consistency that defines these structures should be transitivity and not balance, because the relationships in these relations are not signed.

The salient feature of Tongan social structure is the pervasiveness of rank (Gifford 1929), which characterizes social classes, descent groups, and the ordering of the gods (Williamson 1937), and which is concretely expressed in the *kava* ritual. The hierarchy of the class structure consists of the king (*tu'i*), chiefs (*'eiki*), ceremonial attendants (*matapule*), and commoners (*tu'a*). Politically, descent groups are ranked vis-à-vis each other, and also internally on the basis of genealogical seniority, so that each individual within a group and each subgroup within the larger group is, ideally, assigned a unique rank. Each individual is, as it were, in a class alone.

Rank also characterizes domestic relations: "Relatively speaking, in every household there are chiefs and commoners. As in the family no two individuals are identical in rank, so in the enlarged family" (Gifford 1929:19). This hierarchy is predictable on ethnological grounds, the political structure of a society. In a chiefdom as opposed to an egalitarian band

Digraphs

or tribe (Service 1962; Sahlins 1968), just as political relations between individuals and groups are based on a system of ranking, so are the kinship relations that are, metaphorically speaking, isomorphic with them. The domestic structure represents a "chiefdom within a chiefdom" in which each individual knows his or her place and as Sahlins puts it, "rehearses kinds of behavior and attitudes necessary to the running of the larger system" (Sahlins 1968:63).

For Tonga we can take the atom of kinship as a convenient unit and an appropriate one since it was one of Lévi-Strauss's original examples of this structure. Data for all six relations are in Gifford (1929:16, 17, 18, 22, 23, 29) and may be summarized in this way: A sister's son outranks a mother's brother. This is the famous *fahu* relation, according to which the former is above the law with respect to the latter. The institution of *fahu* is expressed in privileges such as the appropriation of property and is definitely asymmetrical. A sister outranks her brother and according to Gifford "since sisters rank brothers, a sister's husband must be treated with the same respect as his sister. . . . Woman's brother's wife and wife's brother are inferior to the speaker on account of the superiority of sisters to brothers." The husband is always considered head of the immediate family. In general, "the father's side of the family is the chiefly (*'eiki*) side and the mother's side is the *tu'a* commoner side." The father is *tapu* to the child, as expressed in numerous asymmetrical prohibitions on touching him or his possessions. The mother, however, who is on the commoner side of the family, is not *tapu* to the child. The first-born son is definitely superior to the mother (Goldman 1970:453).

For purposes of ethnographic contrast and structural comparison, we can define the same kinship group for Truk in Micronesia, a matrilineal rather than a patrilineal chiefdom. This can be done using Goodenough's (1969) well-known Guttman scale analysis of the Trukese *pin me wöön* system, the "duty scale of setting oneself above another in Truk." This scale is defined by six behavioral norms that characterize interaction between ego and alter, conceived in terms of the duties ego owes to alter. The duties include: (a) the use of a special greeting for alter; (b) an avoidance of being physically higher than alter; (c) an avoidance of initiating interaction with alter; (d) an obligation to honor alter's requests; (e) a prohibition on scolding or criticizing alter for his actions; (f) an avoidance of physical or verbal assault on alter. Using the same abbreviations for kin as in Fig. 3.8 in Chapter 3, status in the Trukese atom is defined as follows:

1. the U/S relation: S outranks U, since S owes U no duties and U owes S d, e, f.
2. the U/M relation: U outranks M, since U owes M no duties and M owes U b, c, d, e, f.
3. the F/U relation: U outranks F, since U owes F no duties and F owes U b, c, d, e, f.

4. the *F/M* relation: neither outranks the other, since *F* and *M* do not owe each other any duties.
5. the *F/S* relation: *S* outranks *F*, since *F* owes *S* no duties and *F* owes *S d, e, f*.
6. the *M/S* relation: *S* outranks *M*, since *S* owes *M* no duties and *M* owes *S f*.

The first digraphs D_1 and D_2 in Fig. 4.16 show the Tongan and Trukese atom of kinship constructed from the data just given, using the same format as in Chapter 3, except that here there are arcs that represent the relation "*u* outranks *v*."

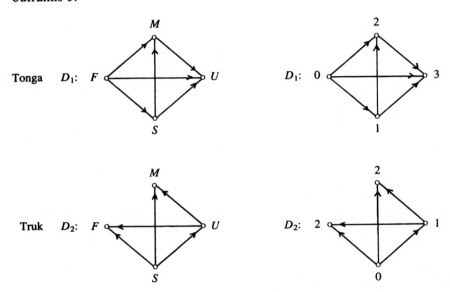

Fig. 4.16. Digraphs and level assignments of Tongan and Trukese kinship relations.

The digraphs D_1 and D_2 in Fig. 4.16 are $\bar{r}\ \bar{s}\ t$ and are therefore partial orders. Empirically, a person does not outrank himself, two persons do not outrank each other, and for any three persons, if *A* outranks *B*, and *B* outranks *C*, then *A* outranks *C*. The ranking is therefore consistent. The digraph D_1 is also *c* and is therefore a complete order or a transitive tournament. The hierarchical structure of these digraphs, and more generally of any acyclic digraph, which may or may not be partial order, can be defined using the concept of level assignment.

An *acyclic digraph D* has no cycles; it is obviously asymmetric, since it has no cycles of length 2. A digraph *D* is said to have a *level assignment*, which assigns a positive integer *n* for each point *v*, called its level, if for each arc $v_i\ v_j$ of *D*, the corresponding integers satisfy $n_i < n_j$. Thus each arc of *D* is directed from a lower to a higher level. If a digraph *D* has a cycle, then it cannot have a level assignment. Empirically, in a level assignment, the lower the integer, the higher the rank (similar to a golf score). The second digraphs in Fig. 4.16 show D_1 and D_2 with level assignments. In D_1 the kinship categories are ranked as $F>S>M>U$, and in D_2 they are ranked as $S>U>F$ and M.

In a complete order, the relative level of every pair of points is determined by D, but in a partial order, it may not be. In D_2, F and M could be assigned different relative levels. For example, if $S=0$ and $U=1$, then $F=2$, $M=3$, and $F=3$, $M=2$ are both possible. Because in this case there is no empirical basis for choosing one assignment over the other, the method adopted is to assign them the same level, interpreted as the same relative rank in a level assignment using the smallest possible number of levels. This number is given by the following theorem (from Harary et al. 1965).

Theorem 4. Let D be an acyclic digraph and let n be the length of the longest path of D. Then $n+1$ is the smallest number of levels in any level assignment of D.

The method of level assignment for an acyclic digraph of any size is simple. A *transmitter* of a digraph is a point whose indegree is 0 and whose outdegree is positive. A *detour* from u to v is a longest path. Let U be the set of transmitters of D and assign the integer 0 to each point of U. To every other point assign the length of a detour to it from some transmitter. Let $v_i v_j$ be an arc of D and let L be any detour to v_i from a transmitter u. Then v_j is not in L; for if it were, the part of path L from v_j to v_i, together with arc $v_i v_j$, would form a cycle contradicting the hypothesis that D is acyclic. Thus v_j has a higher assignment than v_i. This level assignment has precisely $n+1$ levels, since the longest path has length n.

To illustrate, in D_1 in Fig. 4.16, the set U consists of F, which is assigned 0. The points S, M, U are assigned the integers 1, 2, 3, respectively, which is the length of the longest path from F to each of them. The method of level assignment provides an unequivocal, easily applicable definition of relative rank, and it permits one to compare the degree of hierarchy in different social or cultural systems.

The idea of level assignment can be generalized from acyclic digraphs by means of a quasi-level assignment. A *quasi-level assignment* of a digraph D associates with each point v_i a positive integer n_i, called its *quasi-level*, such that: (1) If $v_i v_j$ is an arc of D, then $n_i \leq n_j$ and (2) $n_i = n_j$ if and only if v_i and v_j are mutually reachable. Two points are mutually reachable if and only if they are in the same strong component of D. The concept of quasi-level therefore makes use of the condensation D^* of a digraph D into its strong components.

An effective way of gaining insight into the structural properties of a digraph D is to construct a simple one from it by replacing certain subgraphs of *D* by points, and joining the new points by arcs induced from those of *D* in a specified manner. There are various ways in which this may be done. One is to condense *D*. Consider the set of points *V* of a digraph *D* and let *V* be partitioned into subsets $S_1, S_2, \ldots, S_n$. The *condensation of D with respect to this partition* is the digraph whose points are these *n* subsets (each point being labeled by the symbol used for its corresponding subset), and whose arcs are determined as follows: There is an arc from point S_i to point S_j in the new digraph if and only if in *D* there is at least one arc from a point of S_i to one of S_j. The *condensation D^* of a digraph D into its strong components* takes S_i as the strong components of *D*. It is illustrated in Fig. 4.17. The condensation D^* of *D* is always acyclic and therefore has a level assignment. In a quasi-level assignment, we assign to each point v_i of *D* the level n_i of its strong component in D^*, as we will now illustrate using an ethnographically instructive example from Melanesia.

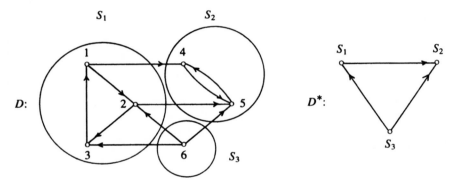

Fig. 4.17. The condensation of a digraph into its strong components.

In most Melanesian societies, rank is not formally defined as it commonly is in Polynesian and Micronesian chiefdoms (at least in the more "highly developed" ones). Instead of a formal, theoretically complete asymmetrical hierarchy defined, say, by birth order, there may be an informal hierarchy containing symmetrical as well as asymmetrical relations, based on achievement of various kinds – for example in oratory, warfare, and especially competitive feasting. Hierarchical elaboration varies considerably, from the rudimentary, such as that of the Arapesh (Mead 1963) to the complex, such as that of Mt. Hagen (Strathern 1971).

In one of the most methodologically self-concious of Melanesian ethnographies, Douglas Oliver (1955) describes rank or stratification in the men's societies of Siuai in the Solomon Islands. These societies are headed by a big man, a *mumi,* and function preeminently in the organization of large-scale feasts. Oliver found that informants could readily define the stratification of such a group, as when they said:

84

Digraphs

Men are either at the top or at the bottom or in the middle. If a man is a *mumi*, he is really on top; if he is a *mouhe* [literally, a roof plate which supports the thatch of a club house], he is not quite so high; if a *turaturana* [courier] or a *mono* [leg], he is at the bottom. A mumi's *pokonopo*, they are in the middle. (Oliver 1955:330)

Unfortunately, Oliver found that this native construct did not define a hierarchical order. The term *pokonopo*, which means "friend," was used by adherents to designate a *mumi*, and the terms *mouhe* and *turaturana* could be used interchangeably. Oliver was therefore required to spend a considerable amount of time observing interaction among all the members of a men's society, in particular observing who gave orders to whom. He eventually discovered four well-defined levels:

Class I. At the top, a leader who, in events having to do with activities of the society, only originates orders.
Class II. Immediately below the leader, one or more subleaders, men of some renown, who receive orders only from the leader, but who originate them to all other members of the society.
Class III. The majority of the society's membership, who receive orders from Classes I and II and occasionally originate orders to each other and to Class IV.
Class IV. One or more members who usually only receive orders, very rarely originating them. (Oliver 1955:404)

These levels correspond precisely to a quasi-level assignment. Fig. 4.18 shows a simplified version of a digraph D such as one might construct in the field in this situation, together with its condensation D^* and a resulting quasi-level assignment to the points of D.

Semilattices

The condensation D^* of the digraph D in Fig. 4.18 is a semilattice, a structure that has been proposed as a general model of formal organizations (see Friedell 1967 and the discussion in Doreian 1971). It ensures a central authority and a least common supervisor for each pair of positions. The concept of a semilattice may be most easily explained by using graph theory. Consider the four types of structure in Fig. 4.19: lattice, semilattice, tree, and oriented graph (asymmetric digraph).

In digraph terms, an *upper bound w* of two points u and v in D is a point that can reach both u and v by directed paths. Further, w is called a *least upper bound (LUB)* for u and v if it is not only an upper bound, but for any other upper bound w', every path from w' to either u or v must contain w. It is an elementary theorem that if w is a least upper bound for u and v, then there can be no other LUB; hence w can be called the LUB for u and v. The directional dual of the LUB (u, v) is their *greatest lower bound (GLB)*, defined similarly.

A *lattice* is a poset in which every two points have both a least upper

85

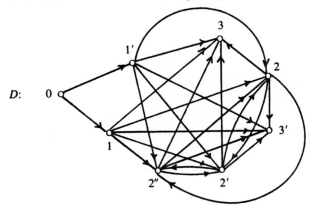

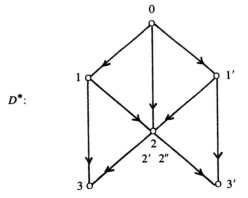

Fig. 4.18. A digraph of stratification in a Siuai big man's society. This D^* is an upper semilattice.

bound and a greatest lower bound. There are some posets in which the LUB of every pair of points exists. These are technically called upper semilattices. As expected, a poset with the property that every two points have a GLB is known as a lower semilattice. However, the first of these is usually taken as the definition of a semilattice. Thus, a *semilattice* is a poset in which every two points have a least upper bound.

We have seen that a tree is a connected graph with no cycles. The concept of a *rooted tree* has already been mentioned in Chapter 2; it is obtained from a tree T by distinguishing one of the points from the others and calling it the *root*. This is usually shown by drawing a small circle around the root point u, but it can also be depicted by converting T into a digraph, as follows. Begin by *orienting* each (undirected) line incident with u away from u

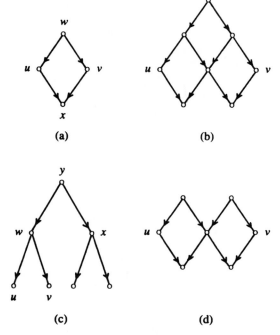

Fig. 4.19. (a) Lattice; (b) semilattice; (c) tree; (d) oriented graph.

by putting an arrow on it. Then continue until every line of T is oriented away from u. The fact that the resulting digraph D is interchangeable with the tree T rooted at u is seen from the construction of u as the unique transmitter of D. This digraph D is sometimes called an *oriented tree from u*, or more briefly an *out-tree*. In general, an *oriented graph* is an asymmetric digraph, so that it satisfies $\bar{r}$ and $\bar{s}$.

Clearly, an oriented tree from a point is a special kind of semilattice in which any two upper bounds of a point pair are comparable; that is, one must be an upper bound of the other. Every lattice and every rooted tree is a semilattice, but not conversely. Thus a semilattice is a more general mathematical structure than a lattice or a tree from a point.

It can be seen that the LUB of points u and v in Fig. 4.19 is w because a directed path exists from w to u and from w to v, making it an upper bound. Further, every other point that can reach both u and v can also reach w, making it a least upper bound. In the out-tree of Fig. 4.19, the LUB of points u and x is point y. Points u and v have two upper bounds w and y, one of which is an upper bound of the other.

On the other hand, the GLB of the two points u and v is defined analogously by directional duality (see Chapter 6); that is, the consideration resulting from reversing the directions of the arcs. For example, point x of Fig. 4.19(a) is the GLB of u and v, but in Fig. 4.19(b), the points u and v do not have a GLB. The lack of a GLB in this latter instance makes it an upper semilattice, since every pair of points does have an LUB. If this digraph were inverted, it would then become a lower semilattice.

In the oriented graph of Fig. 4.19(d), the points u and v have neither an LUB nor a GLB. Hence it is neither a lattice nor a semilattice.

Lattices have been evoked in studies of Melanesian social structure, for example in Kelly's (1974) model of Etoro marriage exchange and interlineage ties, which resembles the street plan of Manhattan, and in Freedman's (1970) model of cooperating "interdigited" kindreds on Siassi. However, these are lattices in a metaphorical sense only.

In organization theory, a contrast is sometimes made between an out-tree (often called just a tree in the literature) and a semilattice that is not an out-tree. One can imagine situations in which one of these structures would be more adaptive than the other. For example,

If in constructing an organization the organizational elite was concerned to minimize conflicting orders, then a tree structure would serve this purpose. Every member of the organization (except the head of the organization) would have one and only one superordinate. There would be no possibility of conflicting orders as each individual has only one source from which orders could be received. (Doreian 1971:99)

A semilattice that is not a tree, on the other hand, allows for the possibility of conflicting orders but is less vulnerable than a tree, in which the flow of communication can be threatened at every line (and at every cutpoint). The multiple channels of communication in the semilattice are also said to "encourage motivation to participate" in the system.

With this distinction in mind, it is interesting to compare the Melanesian system of stratification in Fig. 4.18, the semilattice D^*, which is based on the giving of orders for the frequent organization of large-scale feasts, with the Micronesian system shown in Fig. 4.20.

Fig. 4.20 represents the Yapese chain of authority (Lessa 1950, 1966), an interisland political structure in the Western Carolines. The root point of the tree is the Gagil district on Yap, a large volcanic island that "owns" the 14 small coral islands to its east. Every year, or sometimes every few years, Yap calls for tribute, *pitigil tamol*, which consists of fine woven mats and cloths, sennit, coconut oil, and various other items. The order must be transmitted from island to island in exactly the way shown in this tree; if it is not, then it is ignored. Each of the islands is also subdivided into ranked districts and clans, and on Ulithi at least, there is the same sort of tree for the internal dissemination of orders.

Two observations may be made. First, the tree structure guarantees that the orders of an elite group are transmitted without contradiction and therefore readily accepted. The apparent structural vulnerability – there are no

Digraphs

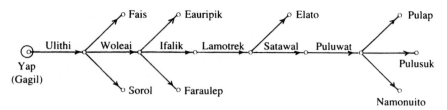

Fig. 4.20. The Yapese chain of authority (from Lessa 1966).

physical sanctions to ensure compliance with orders – is negated by the strong motiviation to participate: The low islands gain the economic help of Yap in times of physical disaster (cyclones sweep the area, destroying crops and land) (Alkire 1977), and Yap gains highly prized "exotic" goods with which to operate its own system of alliances (Lingenfelter 1975; Labby 1976). One might contrast this state of affairs with the type of semilattice in Fig. 4.18, which provides structural motivation to participate; that is, the kind of arm-twisting, manipulating, and cajolery that is often necessary for a Melanesian big man to throw a feast.

Second, the tree structure serves to reinforce a pattern of overlapping authority. According to Lessa, the Yapese are superior to Ulithi and all other islands, metaphorically as fathers to children, and the Ulithians are superior, again as fathers, to the remaining islands (which are not therefore grandchildren of the Yapese). It appears from other sources that there is additional internal ranking, since Lamotrek is superior to Elato and Satawal, from which it receives tribute (Alkire 1965), and Puluwat is superior to Pulap, Pulusuk, and Namonuito (Gladwin 1970). Ulithi, Lamotrek, and Puluwat are all cutpoints in this chain of command – points that pass on orders. Our conjecture is that all the other cutpoints were once in positions of superior rank and that their rank relative to each other was defined by their distance from the rooted point, Yap.

Cyclic structures

Cyclical relations may be implicated in classification as well as in exchange and status systems, and sometimes provide models for the interpretation of both natural and social processes. In Chinese philosophy, the Five Element Theory (*wu hsing*) was based on both a linear order that defined the successive appearance of the elements and two cyclical orders that defined their interaction. The latter are shown in the digraphs (directed cycles) from Joseph Needham's *Science and civilisation in China* (Fig. 4.21).

The elements in each digraph are W = wood, F = fire, E = earth, M = metal, and w = water. In the Mutual Production Order (*hsiang sêng*), the arcs represent the production of one element by another.

89

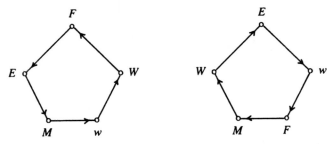

Fig. 4.21. Cyclical structures in the Five Element Theory (from Needham 1956).

W produces F (by being consumed as fuel), F produces E (by giving rise to ashes),[3] E produces M (by fostering the growth of metallic ores within its rocks), M produces w (by attracting or secreting sacred dew when metal mirrors were exposed at night, or else by its property of liquifying), and w produces W (by entering into the substance of plants), thus completing the cycle. (Needham 1956:255)

In the Mutual Conquest Order (*hsiang shêng*), the arcs represent the dominance of one element by another:

Starting with the last link in the cycle, W conquers E (because presumably, it can, when in the form of spades, dig it up and make shapes of it), M conquers W (because it can cut it and carve it), F conquers M (because it can melt and even volatilise it), w conquers F (because it can extinguish it) and E conquers w (because it can dam it up and constrain it – a metaphor very natural for a people whose life depended so much upon hydraulic engineering and irrigation as did that of the Chinese). (Needham 1956:256)

Both orders extended the interpretation of nature by allowing the deduction from them of two secondary principles that are based on graphical distance. The first, the Principle of Control (*hsiang chih*) asserts that "a given process of destruction is 'controlled' by the element which destroys the destroyer" (Needham 1956:257). Thus *W* destroys *E*, but *M* controls the process (by cutting wood faster than wood can dig the earth); *M* destroys *W*, but *F* controls the process; and so on around the cycle. The controlling element is always adjacent to the destroying element.

The second principle, the Principle of Masking (*hsing hua*) "refers to the masking of a process of change by some other process which produces more of the substrate or produces it faster than it can be destroyed by the primary process" (Needham 1956:258). Thus *M* destroys *W*, but *w* "masks" the process by producing wood faster than metal can cut it up. In the Mutual

[3]This may refer to the burning off of vegetation preparatory to cultivation.

Digraphs

Conquest digraph, the masking element is two steps from the destroying element.

Needham emphasizes that these two principles, which were not restricted to naturalistic observation (the Principle of Control, for example, was especially important in "fate-calculation") are perfectly logical in character. Both types occur in such modern sciences as enzyme kinetics and ecology. From the latter, Needham uses the example of the food chain.[4]

The "food chains" in natural oecological communities must obviously depend on the relative abundance of the various species which prey on one another in a sequence based on their sizes and habits. A factor which increases the abundance of a certain bird will indirectly benefit a population of aphids because of the thinning effect which it will have on the coccinnellid beetles ("lady birds") which eat the aphids but are themselves eaten by the birds. Modern economic entomology is full of such examples. (Needham 1956:258)

To illustrate the Principle of Masking:

Larger carnivores may devour the lemmings of Norway but although they may continue to do so at maximal speed, their efforts will rapidly be overtaken in the years when those still mysterious factors operate which so enormously increase the lemming population. Examples of competing processes which would illustrate this quite simple but perfectly justified deduction from the dual cycles of production and destruction respectively, could certainly be found from every branch of modern science. (Needham 1956:259)

The sociological interpretations of these structures are equally interesting. The Mutual Conquest Order provided a historical model explaining the rise and fall of ruling houses as a natural process. (This gave the Naturalists who developed the theory in the third and fourth centuries B.C. considerable influence, since the feudal lords who interpreted it politically were concerned to know under what element they ruled and what steps they might take in the face of inevitable change.) And the Principle of Control provided a "proof" for correct kinship behavior. When this principle is related to both the Mutual Production and Mutual Conquest orders, the "controlling element is always that one produced by the destroyed element" (Needham 1956:258). For example, F destroys M but is controlled by w, but M produces w, giving a 3-cycle containing one productive and two destructive arcs. By analogy, in Confucian interpretation, this demonstated the son's right to take revenge on his father's enemy.

[4]As a biochemical example of the Principle of Control, "... there is an enzyme, phosphorylase, which breaks down glycogen to hexose molecules, esterifying them with phosphate as it does so. But there is another enzyme which breaks the phosphorylase into two portions, thereby inactivating it and making its enzymic function impossible [Keller and Cori]" (Needham 1956:258). Perhaps the first explicit use of digraphs as a mathematical model for food webs is given in Harary (1961d), in which the points are animal species or vegetation and an arc from u to v indicates that species u eats species v. For a graph theoretic analysis of food webs as studied in population biology, see Cohen (1978).

Connectedness

Some time ago, Barnes (1959) reviewed and clarified from a graph theoretic point of view various interpretations of connectedness and connectivity in the social sciences. These terms are sometimes distinguished and sometimes equated and may refer to such diverse properties as the length, the existence, the number, and the proportions of possible paths between pairs of points in a graph or digraph. The two terms are, however, *always* distinguished in graph theory, where connectivity gives a measure of the amount of connectedness. The connectedness of a digraph, as defined earlier, refers to an underlying structure of reaching and joining. It does not refer to density, the ratio of actual to possible lines in a graph, as in Bott's (1971) well-known study of close-knit and loose-knit social networks. One might say that the criterion of a matrilateral connubium is a strong (asymmetric) digraph (Livingstone 1969), but not that it is a dense digraph. The connectivity of a graph is defined in Chapter 7.

Digraphs are a natural model for the analysis of social structure, and they provide the most intuitive conceptualization of the theory of relations, which is fundamental to all areas of anthropology. A graph, as noted earlier, can be characterized in a number of ways – geometrically, topologically, algebraically, and set theoretically. The complexity of the graphs often found in anthropology requires that we go on to consider the algebraic or matrix representation.

5

Graphs and matrices

The method of reducing information, if possible, into charts or synoptic tables ought to be extended to the study of practically all aspects of native life.

Bronislaw Malinowski, *Argonauts of the Western Pacific*

Graphs represented as matrices are not unknown in anthropology. C. G. Seligman, for example, in his early monograph, *The Melanesians of British New Guinea,* summarized social relations between Koita tribal sections in matrix form: "[Table 5.1] shows the sections which were more or less constantly and reciprocally hostile, a cross, where horizontal and vertical lines meet, indicates that the sections referred to would be often at enmity" (Seligman 1910:42). This is the adjacency matrix of a signed graph, except that in conventional notation, the xs would be shown by -1s and the blanks, which presumably signify friendship, by $+1$s, with diagonal entries 0.

Matrix operations are also not unknown, certain of them having been dis-

Table 5.1.

	Gorobe	Badili	Yarogaha	Yawai	Hohodai	Guriu	Baruni	Huhunamo	Roko	Idu	Gevana	Arauwa	Rokurokuna
Gorobe								x		x	x		
Badili								x		x	x		
Yarogaha								x		x	x		
Yawai								x		x	x		
Hohodai								x		x	x		
Guriu								x		x	x		
Baruni								x		x	x		
Huhunamo	x	x	x	x	x	x	x						
Roko													
Idu	x	x	x	x	x	x	x						
Gevana	x	x	x	x	x	x	x						
Arauwa													
Rokurokuna													

covered independently by Yves Lemaître (1970). In an article in the *Société des Océanistes,* he shows how the matrix operations of the product AB, the transpose A', and elementwise multiplication $A \times B$, together with boolean addition $(1 + 1 = 1)$, can be used to find the equivalence classes in a network of interisland voyaging. In the language of graph theory, he develops a method for finding the strong components in a digraph whose points represent islands and whose arcs represent the possibility of sailing directly from one island to another. Thus his matrix A, which shows the 1-step connections between pairs of islands, and his matrix S^n, which shows the n-step connections, correspond to the adjacency and reachability matrices, A and R, of a graph or digraph as defined in Harary et al. (1965, Chapter 5).

We can illustrate these two matrices and provide a purely intuitive introduction to the topic by using the digraph in Fig. 5.1, which is based on Levison, Ward, and Webb's (1973) computer simulation of drift voyaging in Polynesia. The points represent islands or island groups, and the arcs represent a certain probability of drifting, traveling without intent, from one island to another.

In the adjacency matrix $A(D)$, each island has a row and a column, and the entries are 1 if island j can be reached directly from island i, and 0 if not. In the reachability matrix $R(D)$, the entries are 1 if island j can be reached either directly or indirectly (by island hopping) from island i, and 0 if not. (By convention, each point in R can reach itself.) Thus $A(D)$ shows that it would not be possible to drift directly from the Marquesas to the Society Islands, but $R(D)$ shows that it could be done by a succession of drifts.

The adjacency matrix does not support the hypothesis that Polynesia was settled by drift voyaging because it contains transmitters, or islands that cannot be reached from any others, shown by columns with all 0s. The reachability matrix shows that even if one of the transmitters, the Marquesas or Pitcairn, were somehow reached, all the remaining islands could not have been settled from either of them, since the row of neither has all 1s. Thus it becomes necessary to assume at least some voyaging with intent to account for the settlement of Polynesia. As graphs increase in size, it becomes not only easier but eventually necessary to read off structural properties from various of their matrices. Finding the partial betweenness values of all the points in Fig. 2.12, for example, can be done by visual search rather than by using the distance matrix, but not quickly and not with complete confidence in the results.

In this algebraic presentation of graphs, we define five basic matrix operations: addition, subtraction, multiplication, transposition, and the elementwise product. We also define three of the matrices associated with a graph: the adjacency, reachability, and distance matrices.[1] We emphasize

---

[1] Some of the mathematical concepts of this chapter are based on Chapter 5 of Harary et al. 1965 and on Harary 1969.

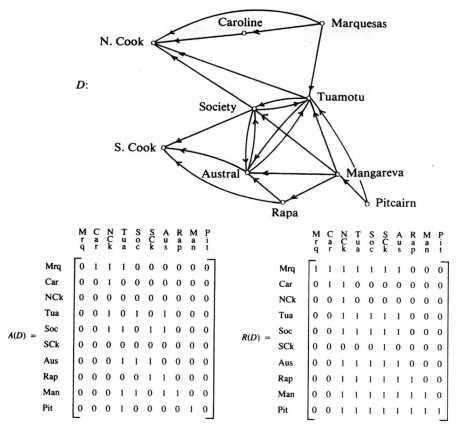

$$A(D) =$$

	Mrq	Car	NCk	Tua	Soc	SCk	Aus	Rap	Man	Pit
Mrq	0	1	1	1	0	0	0	0	0	0
Car	0	0	1	0	0	0	0	0	0	0
NCk	0	0	0	0	0	0	0	0	0	0
Tua	0	0	1	0	1	0	1	0	0	0
Soc	0	0	1	1	0	1	1	0	0	0
SCk	0	0	0	0	0	0	0	0	0	0
Aus	0	0	0	1	1	1	0	0	0	0
Rap	0	0	0	0	0	1	1	0	0	0
Man	0	0	0	1	1	0	1	1	0	0
Pit	0	0	0	1	0	0	0	0	1	0

$$R(D) =$$

	Mrq	Car	NCk	Tua	Soc	SCk	Aus	Rap	Man	Pit
Mrq	1	1	1	1	1	1	1	0	0	0
Car	0	1	1	0	0	0	0	0	0	0
NCk	0	0	1	0	0	0	0	0	0	0
Tua	0	0	1	1	1	1	1	0	0	0
Soc	0	0	1	1	1	1	1	0	0	0
SCk	0	0	0	0	0	1	0	0	0	0
Aus	0	0	1	1	1	1	1	0	0	0
Rap	0	0	1	1	1	1	1	1	0	0
Man	0	0	1	1	1	1	1	1	1	0
Pit	0	0	1	1	1	1	1	1	1	1

Fig. 5.1. An illustration of the adjacency and reachability matrices of a digraph of drift voyaging in eastern Polynesia (digraph adapted from the network in Levison, Ward, and Webb 1973).

the significance of the adjacency matrix as both an "ethnographic instrument" and an "ethnological document," to use Malinowski's (1961) characterization of his famous "mental charts," and we indicate how this matrix reflects some of the basic structural properties of a graph. The ethnographic examples include rank and intercaste food transactions as analyzed by Marriott (1968), and status and communication in the Melanesian big man system, analyzed in Chapter 2.

The adjacency matrix

There are numerous ways to represent a graph, all of which are equivalent in the sense that they define the same structure. One can use diagrams, as we have done so far; one can also list lines (lines or arcs) or use "graphical numbers" or labeled bipartite graphs. There are, in addition, matrix representations. These include the adjacency matrix, which shows the adjacency of points; the incidence matrix, which shows the incidence of points with lines; the line adjacency matrix, which shows the adjacency of lines; and the distance matrix, which shows the distance between every pair of points and therefore their adjacency as well. When a graph is small, a diagram is usually superior to a list, because it permits immediate comprehension of a structure. When a graph is large, however, a matrix representation is superior, and it is necessary for the algebraic manipulations that provide insight into and quantification of structural properties. From an anthropological point of view, the most important and natural of the matrix representations is the adjacency matrix.

The *adjacency matrix of a labeled graph* G with p points written $A = A(G) = [a_{ij}]$ is the $p \times p$ matrix in which $a_{ij} = 1$ if v_i is adjacent with v_j and $a_{ij} = 0$ otherwise. Thus A is a *binary matrix* because each entry is 0 or 1, and there is a one-to-one correspondence between labeled graphs with p points and $p \times p$ symmetric binary matrices with zero diagonal. Because of this correspondence, every graph theoretic concept is reflected in the adjacency matrix. Thus, for example, the row (or equivalently, the column) sums of A are the degrees of the points of G, as can be seen in Fig. 5.2, which shows a graph and its adjacency matrix. The numbers 1, 2, 3, 4 to the left of the rows and above the columns indicate the points.

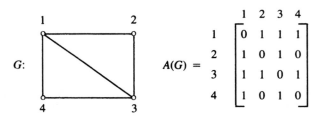

Fig. 5.2. A graph and its adjacency matrix.

The *adjacency matrix of a labeled digraph* D is defined similarly: $A = A(D) = [a_{ij}]$ has $a_{ij} = 1$ if arc v_iv_j is in D and 0 otherwise. Note that $A(D)$ is not necessarily symmetric. In $A(D)$ the row sums give the outdegrees and the column sums the indegrees of the points, as can be seen in Fig. 5.3.

In the *adjacency matrix of a labeled signed graph or digraph* S, $A = A(S) = [a_{ij}]$, the entries are $a_{ij} = +1$ or -1 depending on the sign of the line or arc, and $a_{ij} = 0$ if there is no line or arc, as illustrated in Fig. 5.4.

Since there is a one-to-one correspondence between a labeled graph and its adjacency matrix, two graphs are certainly isomorphic if their adjacency matrices are equal. Two matrices $A = [a_{ij}]$ and $B = [b_{ij}]$ are *equal matrices* if

Graphs and matrices

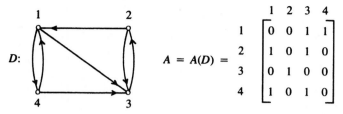

Fig. 5.3. A digraph and its adjacency matrix.

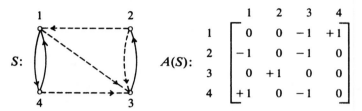

Fig. 5.4. A signed digraph and its adjacency matrix.

$a_{ij} = b_{ij}$ for all i and j. This is illustrated in Fig. 5.5 for two "different look-ing" graphs G_1 and G_2, in which the points of G_2 have been assigned in such a way as to show its isomorphism with G_1 by means of equal adjacency matrices.

We can use the graphs in Fig. 5.5 to show one way in which the patterning of an adjacency matrix reflects the structural properties of a graph. Graphs of this type are known in mathematics as bipartite graphs, and they implicitly occur in anthropology as an underlying structure of dual organization; that is, the division of a society into moieties or halves for purposes of marriage, competitive feasting, ritual, and so on.

A *bipartite graph* (or *bigraph*) G is a graph whose point set V can be partitioned into two subsets V_1 and V_2 such that every line of G joins V_1 with V_2. Such a graph is also called bicolorable (see Chapter 2). In Fig. 5.5, G_1 and G_2 are isomorphic so that each has as its subsets $V_1 = \{1, 3, 5, 7, 9\}$ and $V_2 = \{2, 4, 6, 8\}$.

It happens sometimes that a society has unrecognized dual divisions – for example, the relations of enmity and friendship as among the Koita or relations of marriage exchange as among the Etoro (Kelly 1974) – both of which systems are depicted by implicit adjacency matrices arranged in such a way as to emphasize this structure. The bipartite structure of a large graph can be shown by partitioning its adjacency matrix in a certain way.

Given an adjacency matrix A, we separate its set of rows into two subsets, the first r rows and the rest. We also separate its columns into the first s columns and the rest. Thus we obtain a *partitioned matrix*:

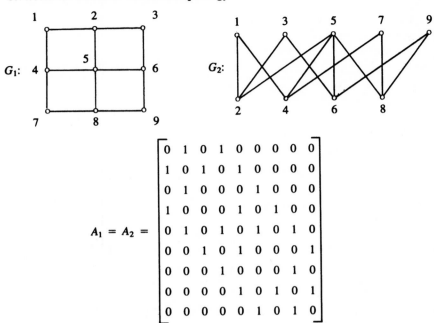

$$A_1 = A_2 = \begin{bmatrix} 0 & 1 & 0 & 1 & 0 & 0 & 0 & 0 & 0 \\ 1 & 0 & 1 & 0 & 1 & 0 & 0 & 0 & 0 \\ 0 & 1 & 0 & 0 & 0 & 1 & 0 & 0 & 0 \\ 1 & 0 & 0 & 0 & 1 & 0 & 1 & 0 & 0 \\ 0 & 1 & 0 & 1 & 0 & 1 & 0 & 1 & 0 \\ 0 & 0 & 1 & 0 & 1 & 0 & 0 & 0 & 1 \\ 0 & 0 & 0 & 1 & 0 & 0 & 0 & 1 & 0 \\ 0 & 0 & 0 & 0 & 1 & 0 & 1 & 0 & 1 \\ 0 & 0 & 0 & 0 & 0 & 1 & 0 & 1 & 0 \end{bmatrix}$$

Fig. 5.5. Two isomorphic graphs and their common adjacency matrix.

$$A \;=\; \begin{bmatrix} A_{11} & A_{12} \\ A_{21} & A_{22} \end{bmatrix}.$$

If the symmetric matrix A can be partitioned into submatrices such that A_{11} and A_{22} are square and consist entirely of zeros, then G is bipartite, as illustrated for the adjacency matrix of G_1 in Fig. 5.5.

	1	3	5	7	9	2	4	6	8
1	0	0	0	0	0	1	1	0	0
3	0	0	0	0	0	1	0	1	0
5	0	0	0	0	0	1	1	1	1
7	0	0	0	0	0	0	1	0	1
9	0	0	0	0	0	0	0	1	1
2	1	1	1	0	0	0	0	0	0
4	1	0	1	1	0	0	0	0	0
6	0	1	1	0	1	0	0	0	0
8	0	0	1	1	1	0	0	0	0

Graphs and matrices

We call attention here to a related type of partition: If A can be partitioned into submatrices such that A_{11} and A_{22} are square and A_{12} and A_{21} consist entirely of zeros, then A is said to be *decomposed* by this partition and the graph is disconnected, as illustrated in Fig. 5.6.

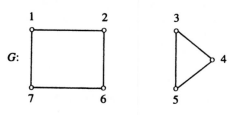

One way in which the adjacency matrix of a digraph can be rearranged to reveal structural properties of a graph is to see if it is *upper triangular*; that is, to see if the rows and columns can be reordered such that all the 1s are above the main diagonal. An example is the adjacency matrix of the digraph of the Tongan atom of kinship depicted in Fig. 4.15. Since it is upper triangular, we know that the properties in Theorem 5.1 must be true of the digraph.

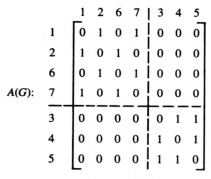

Fig. 5.6. The decomposed adjacency matrix of a disconnected graph.

$$A(D_1) = \begin{array}{c} \\ F \\ S \\ M \\ U \end{array} \begin{array}{cccc} F & S & M & U \\ \left[\begin{array}{cccc} 0 & 1 & 1 & 1 \\ 0 & 0 & 1 & 1 \\ 0 & 0 & 0 & 1 \\ 0 & 0 & 0 & 0 \end{array}\right] \end{array}$$

Theorem 5.1. The following properties of a digraph D are equivalent.

(1) D has no directed cycles.

(2) Every strong component of D is trivial (consists of one point).

(3) The condensation D^* is isomorphic to D.

(4) It is possible to order the points of D so that its adjacency matrix is upper triangular.

(5) It is possible to assign levels n_i to the points v_i in such a way that if $v_i v_j$ is in D, then $n_i < n_j$.

The level assignment for this acyclic digraph was given in Chapter 4. This is, of course, a miniature status system and hardly requires a matrix analysis. For the much larger Indian caste system described below, however, only a matrix analysis is feasible. In this case, we know that none of the properties enumerated in Theorem 5.1 apply, since the adjacency matrices of status relations are not upper triangular (see Fig. 5.8).

In the classic text on fieldwork methods, Malinowski recommends the use of "mental charts" to facilitate complete, thorough, and detailed ethnography. Such charts, he said, "ought to materialize into a diagram, a plan, an exhaustive synoptic table of cases" (Malinowski 1961:14). They should be used in numerous and diverse contexts, for example, economic transactions, ceremonial performances, magical procedures, kinship terms. One can envisage all sorts of diagrams, such as flow charts, tables of correspondence, maps, and certainly graphs like the implicit one he uses to depict the Kula Ring. An explicit use of a graphical representation has the advantage of forcing the ethnographer to define the properties and enumerate the sets of objects being studied, whether they be social relations or cultural categories. The representation of a graph by an adjacency matrix has the further advantage of ensuring an exhaustive description, because one cannot leave out the entry of a cell in a matrix as one can, inadvertently, omit a line in a diagram (as Read does, for example, in his signed graph of Gahuku-Gama alliance structure shown in Chapter 3). The adjacency matrix is also an efficient device for the storage and retrieval of information in the field.

A dramatic illustration of the use of an adjacency matrix is provided by McKim Marriott (1968). In a study of an Indian village, Marriott concluded that the real logic underlying caste rank was precisely what his informants said it was, that relations of superiority and inferiority are established by transactions in food and services: " ... any transfer of food *always* makes the giver higher, the receiver lower, ... any rendering of service *always* makes the master higher, the servant lower" (Marriott 1968:147). Accordingly, he undertook to record a number of culturally defined, distinct types of transactions between all pairs of castes in the village. His grappling with the problem of representation is worth quoting.

After leaving the village, my first attempts to represent the observed food transactions took the form of sociograms like the one for *pakkā* food reproduced in [Fig. 5.7]. A general impression of the preponderantly downward direction of food transfers may be apparent to some viewers from such a sociogram. But just how a precise order of ranking might be derived from this picture is not at all clear to the eye. A long list of graphic conventions about the discovery and representation of rank would have to be adopted and explained; such explanations would reduce the sociogram's apparent initial advantage of immediate intelligibility. Clarity of representation conflicts with full presentation of such data. Finally, my attempt to combine in one sociogram all data on the several kinds of food transactions led me to a tangled multicolored result (not shown) which I could not readily interpret myself, and which even the most expert other viewers found unintelligible.

Graphs and matrices

A suitable, more complete means of representation already widely used and understood in the field of sociometry was suggested to me by Harrison C. White in 1961, as exemplified in the text of Kemeny, Snell and Thompson (1956, pp. 307–315). This is the mathematical matrix of binary numbers. Such a matrix permits a concise, orderly, and exhaustive statement of all data on dichotomous transactions among many participants. The use of binary numbers facilitates rigorous analysis of simple propositions and makes explicit provision for such mathematical manipulations of groups as may later be required. An example is the matrix given in [Fig. 5.8], which represents the same data on *pakkā* food that are represented in the sociogram of [Fig. 5.7]. (Marriott 1968:147)

Fig. 5.7 is a diagram whose points represent castes and whose lines represent the relation "gives pakkā food to." Fig. 5.8 is actually an adjacency matrix of a relation that contains more food transactions than are, or could conveniently be, represented in Fig. 5.7.[2] Marriott has modified the adjacency matrix so that instead of zeros on the diagonal, there are the code numbers of each caste. Thus 1 is the Sanadhya Brahman, 2 is the Maithil carpenter, ..., 36 is the Bhanyi sweeper caste. Normally, as we have seen, the numbers that label the points of a digraph are placed down the side and across the top to identify the rows and columns. To the right of the matrix are the row sums (outdegrees), followed by the column sums (indegrees), and then the difference between them (the "net"). This difference is used to define the relative rank of each caste for this transaction.

Marriott constructs a separate adjacency matrix for each of five food transactions (or food-related transactions) and then manipulates three of these to arrive at a final determination of intercaste rank. One of these manipulations is matrix addition.

Matrix operations

Large graphs can be analyzed by using matrix algebra, and this usually requires the use of computers. Fortunately, computer programs based on the following matrix operations are easy to write and inexpensive to use.

We now define the *sum* $(A + B)$, the *difference* $(A - B)$, and the *elementwise product* $(A \times B)$ of two matrices A and B of the same size $m \times n$. For a matrix C, it is convenient to denote its general (or i,j) entry by $(C)_{ij} = c_{ij}$, as in:

$$(A + B)_{ij} = a_{ij} + b_{ij} \ ,$$

$$(A - B)_{ij} = a_{ij} - b_{ij} \ ,$$

$$(A \times B)_{ij} = a_{ij}b_{ij} \ .$$

---

[2] As complicated as Fig. 5.7 is, it still does not contain all the lines necessary to represent all the food transactions shown in the matrix, as may be seen, for example, by comparing the lines from point 26 with the entries in that row of the matrix.

Structural models in anthropology

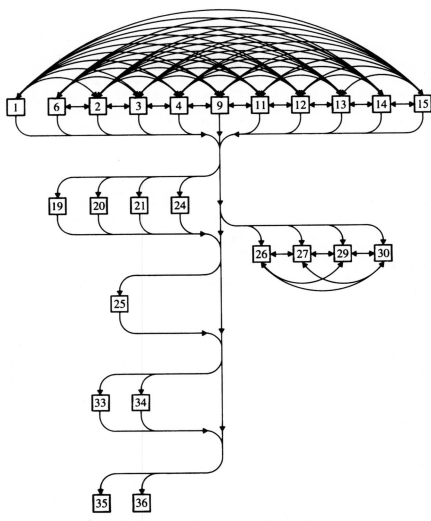

Fig. 5.7. Marriott's (1968) sociogram of intercaste food transactions.

These operations are illustrated by the following matrices:

$$A = \begin{bmatrix} 1 & 0 & 2 \\ 0 & 3 & 0 \end{bmatrix} \qquad A+B = \begin{bmatrix} 1 & 4 & 7 \\ 6 & 10 & 0 \end{bmatrix}$$

$$B = \begin{bmatrix} 0 & 4 & 5 \\ 6 & 7 & 0 \end{bmatrix} \qquad A-B = \begin{bmatrix} 1 & -4 & -3 \\ -6 & -4 & 0 \end{bmatrix}$$

Graphs and matrices

Fig. 5.8. Marriott's (1968) binary matrix representation of intercaste food transactions.

The matrix (Giver Castes as rows, Receiver Castes as columns) with marginal totals Given, Received, Net:

	c1	c2	c3	c4	c5	c6	c7	c8	c9	c10	c11	c12	c13	c14	c15	c16	c17	c18	c19	c20	c21	c22	c23	c24	Given	Received	Net
	(1)	0	1	1	1	1	1	1	1	1	1	1	1	1	1	1	1	1	1	1	1	1	1	1	22	9	13
	0	(6)	1	1	1	1	1	1	1	1	1	1	1	1	1	1	1	1	1	1	1	1	1	1	22	9	13
	1	1	(2)	1	1	1	1	1	1	1	1	1	1	1	1	1	1	1	1	1	1	1	1	1	23	10	13
	1	1	1	(3)	1	1	1	1	1	1	1	1	1	1	1	1	1	1	1	1	1	1	1	1	23	10	13
	1	1	1	1	(4)	1	1	1	1	1	1	1	1	1	1	1	1	1	1	1	1	1	1	1	23	10	13
	1	1	1	1	1	(9)	1	1	1	1	1	1	1	1	1	1	1	1	1	1	1	1	1	1	23	10	13
	1	1	1	1	1	1	(11)	1	1	1	1	1	1	1	1	1	1	1	1	1	1	1	1	1	23	10	13
	1	1	1	1	1	1	1	(12)	1	1	1	1	1	1	1	1	1	1	1	1	1	1	1	1	23	10	13
	1	1	1	1	1	1	1	1	(13)	1	1	1	1	1	1	1	1	1	1	1	1	1	1	1	23	10	13
	1	1	1	1	1	1	1	1	1	(15)	1	1	1	1	1	1	1	1	1	1	1	1	1	1	23	10	13
	1	1	1	1	1	1	1	1	1	1	(14)	1	1	1	1	1	1	1	1	1	1	1	1	1	23	10	13
	0	0	0	0	0	0	0	0	0	0	0	(19)	0	0	0	1	0	0	0	0	1	1	1	1	5	11	− 6
	0	0	0	0	0	0	0	0	0	0	0	0	(20)	0	0	1	0	0	0	0	1	1	1	1	5	11	− 6
	0	0	0	0	0	0	0	0	0	0	0	0	0	(21)	0	1	0	0	0	0	1	1	1	1	5	11	− 6
	0	0	0	0	0	0	0	0	0	0	0	0	0	0	(24)	1	0	0	0	0	1	1	1	1	5	11	− 6
	0	0	0	0	0	0	0	0	0	0	0	0	0	0	0	(25)	0	0	0	0	1	1	1	1	4	15	−11
	0	0	0	0	0	0	0	0	0	0	0	0	0	0	0	0	(26)	1	1	1	0	0	1	1	5	14	− 9
	0	0	0	0	0	0	0	0	0	0	0	0	0	0	0	0	1	(27)	1	1	0	0	1	1	5	14	− 9
	0	0	0	0	0	0	0	0	0	0	0	0	0	0	0	0	1	1	(30)	1	0	0	1	1	5	14	− 9
	0	0	0	0	0	0	0	0	0	0	0	0	0	0	0	0	1	1	1	(29)	0	0	1	1	5	14	− 9
	0	0	0	0	0	0	0	0	0	0	0	0	0	0	0	0	0	0	0	0	(33)	0	1	1	2	16	−14
	0	0	0	0	0	0	0	0	0	0	0	0	0	0	0	0	0	0	0	0	0	(34)	1	1	2	16	−14
	0	0	0	0	0	0	0	0	0	0	0	0	0	0	0	0	0	0	0	0	0	0	(35)	1	1	23	−22
	0	0	0	0	0	0	0	0	0	0	0	0	0	0	0	0	0	0	0	0	0	0	1	(36)	1	23	−22
Totals	9	9	10	10	10	10	10	10	10	10	10	11	11	11	11	15	14	14	14	14	16	16	23	23			

$$A \times B \;=\; \begin{bmatrix} 0 & 0 & 10 \\ 0 & 21 & 0 \end{bmatrix}$$

The operation of multiplication of two matrices A and B is defined only when the number of columns of A equals the number of rows of B. Thus consider A as $m \times n$ and B as $n \times r$. Their *product* $C = AB$ is defined by the formula:

$$c_{ij} = a_{i1}b_{1j} + a_{i2}b_{2j} + \cdots + a_{in}b_{nj} = \sum_{k=1}^{n} a_{ik}b_{kj} \ .$$

103

Thus c_{ij} is obtained from the ith row of A and the jth column of B by what is often called the "inner product" of two vectors.

In particular, the product of a square $n \times n$ matrix A with itself is always defined. Here A is called a *square matrix of order n*. The product AA is written A^2 and its i,j entry is $a_{ij}^{(2)}$:

$$a_{ij}^{(2)} = a_{i1}a_{1j} + a_{i2}a_{2j} + \ldots + a_{in}a_{nj} .$$

Similarly, A^r denotes the rth power of matrix A, and its general entry is written $a_{ij}^{(r)}$. Because the elementwise product $A \times B$ and the product AB are different operations, in general $A \times A \neq A^2$.

It can be shown (Harary 1969:151) that when the adjacency matrix A is raised to higher powers, the entries in A^n show the number of walks of length n from v_i to v_j. Thus A^2 shows the number of walks of length 2 from v_i to v_j, A^3 shows the number of walks of length 3 from v_i to v_j, and so on, as illustrated using the digraph D of Fig. 5.3.

We note here that relative accessibility or centrality is sometimes defined by the number of different ways one place can be reached from other places; that is, by the number of walks of a given length or of all lengths up to the diameter of a graph. See, for example, Taafe and Gautier (1973) on the geography of transportation networks. This particular version of centrality differs from all three versions introduced in Chapter 2.

$$A^2 = \begin{array}{c c} & \begin{array}{cccc} 1 & 2 & 3 & 4 \end{array} \\ \begin{array}{c} 1 \\ 2 \\ 3 \\ 4 \end{array} & \left[\begin{array}{cccc} 1 & 1 & 1 & 0 \\ 0 & 1 & 1 & 1 \\ 1 & 0 & 1 & 0 \\ 0 & 1 & 1 & 1 \end{array} \right] \end{array} \qquad A^3 = \begin{array}{c c} & \begin{array}{cccc} 1 & 2 & 3 & 4 \end{array} \\ \begin{array}{c} 1 \\ 2 \\ 3 \\ 4 \end{array} & \left[\begin{array}{cccc} 1 & 1 & 2 & 1 \\ 2 & 1 & 2 & 0 \\ 0 & 1 & 1 & 1 \\ 2 & 1 & 2 & 0 \end{array} \right] \end{array}$$

We now define the operation A' on matrix A, which corresponds to the converse of a relation or of a digraph. The *transpose A'* of a matrix A is obtained from A by interchanging its rows and columns. Thus the i,j entry of A' is the same as the j,i entry of A. Symbolically, if $A = [a_{ij}]$ and $A' = [a'_{ij}]$, then $a'_{ij} = a_{ji}$. Formally, a matrix is called *symmetric* if it equals its own transpose; that is, $A' = A$. The transpose operation is illustrated in Fig. 5.9. This digraph D' is the converse of D in Fig. 5.3, and its adjacency matrix A' is the transpose of A in Fig. 5.3. Table 5.2 summarizes these five matrix operations.

Two special matrices need to be defined. The *identity matrix* of order n is denoted I_n, or more briefly I, when its order is clear by context. Its entries are 1 on the diagonal and 0 elsewhere. The reason for the name of this matrix is that for any square matrix C having the same order as I, the product $IC = C$, and also $CI = C$. Thus matrix I is the identity matrix with

Graphs and matrices

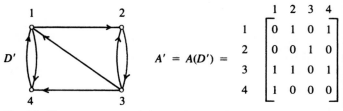

$$A' = A(D') = \begin{array}{c c} & \begin{array}{cccc} 1 & 2 & 3 & 4 \end{array} \\ \begin{array}{c} 1 \\ 2 \\ 3 \\ 4 \end{array} & \begin{bmatrix} 0 & 1 & 0 & 1 \\ 0 & 0 & 1 & 0 \\ 1 & 1 & 0 & 1 \\ 1 & 0 & 0 & 0 \end{bmatrix} \end{array}$$

Fig. 5.9. The transpose A' of the matrix A in Fig. 5.3.

Table 5.2. Matrix operations

Size of A	Size of B	Operation	Notation	Size of result	i, j entry of result
$m \times n$	$m \times n$	Sum	$A + B$	$m \times n$	$a_{ij} + b_{ij}$
$m \times n$	$m \times n$	Difference	$A - B$	$m \times n$	$a_{ij} - b_{ij}$
$m \times n$	$m \times n$	Elementwise product	$A \times B$	$m \times n$	$a_{ij} b_{ij}$
$m \times n$	$n \times r$	Product	AB	$m \times r$	$\displaystyle\sum_{k=1}^{n} a_{ik} b_{kj}$
$m \times n$		Transpose	A'	$n \times m$	a_{ji}

respect to multiplication. The *universal matrix J* is a matrix all of whose entries are 1. The matrix I is always square, whereas J is not necessarily square. Thus, for example,

$$I = I_4 = \begin{bmatrix} 1 & 0 & 0 & 0 \\ 0 & 1 & 0 & 0 \\ 0 & 0 & 1 & 0 \\ 0 & 0 & 0 & 1 \end{bmatrix}$$

$$J = J_{3,4} = \begin{bmatrix} 1 & 1 & 1 & 1 \\ 1 & 1 & 1 & 1 \\ 1 & 1 & 1 & 1 \end{bmatrix}$$

It is also necessary to define *boolean arithmetic* on the integers 0 and 1 for several uses of matrices. With only one exception, addition and multiplication are exactly the same as for ordinary arithmetic. The exception is that $1 + 1 = 1$. It follows that a sequence of ordinary arithmetic operations on 0 and 1 that leads to any positive integer will yield the element 1 using boolean operations. If x and y are any two nonnegative integers, we indicate their boolean sum by the symbol $(x + y)\#$, so that the sum is either 0 or 1. Since $(1 + 1)\# = 1$, we write $2\# = 1$, $3\# = 1$, and so forth. Thus $(2 + 3)\# = 5\# = 1$.

105

Structural models in anthropology

Matrix operations can be used to find various structural properties of a graph, such as the clusterability of a signed graph (Cartwright and Gleason 1967) or the degree of transitivity of a digraph (Harary and Kommel 1979).

The *transitivity ratio* of a digraph D is the probability that if there is a 2-path in D, say from u to v, then the arc uv is also in D. Thus, using picture writing, we have

$$\text{Transitivity ratio of } D = \frac{\text{number of } \triangle}{\text{number of } \wedge} .$$

Both the numerator and the denominator of this ratio are readily expressed in terms of $A = A(D)$. First, for the denominator, we note that $a_{ij}^{(2)}$ is the number of 2-paths from v_i to v_j when $i \neq j$, but that when $i = j$, it is the number of symmetric pairs containing v_i. This gives the sum of all the entries in the matrix $A^2 - \text{diag } A^2$ obtained from A^2 by changing every diagonal entry to 0. It is customary to write this sum as $\Sigma (A^2 - \text{diag } A^2)$.

The numerator of the transitivity ratio is precisely the number of transitive triples in D, each of which results when there is both a 2-path and a 1-path from v_i to v_j. The definition of elementwise product could have been designed for just this purpose, because the i,j entry of $A^2 \times A$ gives the number of transitive triples from v_i to v_j. There is no need to worry about the diagonal as diag $A = 0$, so that diag $(A^2 \times A)$ must also be 0. On summing over all the entries in $A^2 \times A$, we therefore get $\Sigma (A^2 \times A)$, so that

$$\text{Transitivity ratio of } D = \frac{\Sigma (A^2 \times A)}{\Sigma (A^2 - \text{diag } A^2)} .$$

As an example, the transitivity ratio of the digraph in Fig. 5.3 is 0.43.

Addition can be used on two or more adjacency matrices to find the structure of multiplex links in a social network. Marriott, for example, summed his three adjacency matrices representing three distinct food transactions between castes to get a combined matrix just like Fig. 5.8, except that the entries are 0, 1, 2, or 3, depending on whether caste A gives none, one, two, or three types of food to caste B. And multiplication can be used on a single adjacency matrix to find the structure of reachability and distance in a graph.

The reachability matrix

Recall that a point v_j is reachable from a point v_i if there is a path from v_i to v_j. The *reachability matrix* of a digraph D or a graph G is a matrix with entries $r_{ij} = 1$ if v_j is reachable from v_i, and $r_{ij} = 0$ otherwise. Fig. 5.10 shows a digraph and its reachability matrix $R(D)$. It is conventional to stipulate that every point is reachable from itself, as shown by the diagonal entries, $r_{ii} = 1$. This does not contradict the fact that there are no loops in D, but

Graphs and matrices

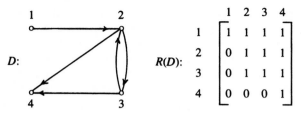

Fig. 5.10. The reachability matrix of a digraph.

rather is based on the following consideration: When the points represent persons each having a certain item of information that is transmitted along the arcs of D, reachability reflects knowledge of that item. As each person already knows his own item, each point must be regarded as reachable from itself. This is the setting for the so-called Telephone Gossip Problem, which asks how many phone calls are necessary in a strong digraph for all persons in D to learn all items of information when at the beginning of the process each person knows a different item; see Harary and Schwenk (1974a,b).

To obtain the reachability matrix R of a graph from its adjacency matrix A, we make use of the following theorems.

Theorem 5.2. For a labeled graph or digraph with adjacency matrix A, the i,j entry of A^n is the number of walks of length n from v_i to v_j.

Since in constructing R we are concerned only with whether v_i can or cannot reach v_j and not with the number of ways it can be done, we make use of boolean addition. By $A^2\#$ we shall mean the matrix obtained by applying boolean arithmetic to compute the entries of the matrix A^2. Symbolically, $A^2\# = [a_{ij}^{(n)}\#]$, and in general by $A^n\#$ we shall mean the matrix obtained when A^n is computed using boolean operations. The next statement is an immediate consequence of a combination of Theorem 5.2 with the definition of boolean arithmetic.

Theorem 5.3. The i,j entry $a_{ij}^{(n)}\#$ of $A^n\#$ is 1 if and only if there is at least one walk of length n in D from v_i to v_j.

This theorem is illustrated using the digraph in Fig. 5.3. The following matrices, in which the rows and columns have the same ordering as the points of the figure, should be compared with the matrices A^2 and A^3.

$$A^2\# = \begin{bmatrix} 1 & 1 & 1 & 0 \\ 0 & 1 & 1 & 1 \\ 1 & 0 & 1 & 0 \\ 0 & 1 & 1 & 1 \end{bmatrix} \qquad A^3\# = \begin{bmatrix} 1 & 1 & 1 & 1 \\ 1 & 1 & 1 & 0 \\ 0 & 1 & 1 & 1 \\ 1 & 1 & 1 & 0 \end{bmatrix}$$

107

By *limited reachability* is meant reachability within a certain number n of steps. Thus R_2 means the reachability matrix for paths of length 2 or less, R_3 for paths of length at most 3, and so on. Let us denote by R the reachability matrix within $p-1$ steps, which is the length of any longest possible path in a graph or digraph with p points. (Note that the two preceding theorems refer to walks, but any walk contains a path.) Then the reachability matrices R_n and R are determined by the next theorem.

Theorem 5.4. For every positive integer n, $R_n = (I + A + A^2 + \cdots + A^n)\# = (I + A)^n\#$ and $R = (I + A + A^2 + \cdots + A^{p-1})\# = (I + A)^{p-1}\#$.

In other words, we get the reachability matrix by adding, in boolean fashion, the identity matrix, the adjacency matrix, and successively higher powers of the adjacency matrix up to power $p-1$.

An important use of the reachability matrix, which is the one independently discovered by Lemaître, is to find the equivalence classes, strong components, in a communication network. This is easily done by using the operations of the transpose and the elementwise product.

Let R be the reachability matrix of D. We form the elementwise product $R \times R'$. Now $r_{ij} = 1$ when v_j is reachable from v_i, and $r'_{ij} = 1$ when v_i is reachable from v_j. Thus the elementwise product $r_{ij} r'_{ij} = 1$ when v_i and v_j are mutually reachable. In the symmetric matrix $R \times R'$ the entries of 1 in the ith row give all the points mutually reachable with v_i.

Theorem 5.5. Let v_i be a point of a digraph D. Then the strong component of D containing v_i is given by the entries of 1 in the ith row (or column) of $R \times R'$.

We illustrate this theorem using the digraph of status in a Siuai big man society, as described in Chapter 4. The matrices R and R' of the digraph in Fig. 4.18 are shown below. The rows of $R \times R'$ show its strong components, which are $\{0\}$, $\{1\}\{1'\}$, $\{2, 2', 2''\}$, $\{3\}\{3'\}$.

	0	1	1′	2	2′	2″	3	3′
0	1	1	1	1	1	1	1	1
1	0	1	0	1	1	1	1	1
1′	0	0	1	1	1	1	1	1
2	0	0	0	1	1	1	1	1
2′	0	0	0	1	1	1	1	1
2″	0	0	0	1	1	1	1	1
3	0	0	0	0	0	0	1	0
3′	0	0	0	0	0	0	0	1

R:

R':

	0	1	1'	2	2'	2"	3	3'
0	1	0	0	0	0	0	0	0
1	1	1	0	0	0	0	0	0
1'	1	0	1	0	0	0	0	0
2	1	1	1	1	1	1	0	0
2'	1	1	1	1	1	1	0	0
2"	1	1	1	1	1	1	0	0
3	1	1	1	1	1	1	1	0
3'	1	1	1	1	1	1	0	1

$R \times R'$:

	0	1	1'	2	2'	2"	3	3'
0	1	0	0	0	0	0	0	0
1	0	1	0	0	0	0	0	0
1'	0	0	1	0	0	0	0	0
2	0	0	0	1	1	1	0	0
2'	0	0	0	1	1	1	0	0
2"	0	0	0	1	1	1	0	0
3	0	0	0	0	0	0	1	0
3'	0	0	0	0	0	0	0	1

The distance matrix

Next to the adjacency matrix, the distance matrix probably has the broadest anthropological applications. It is defined as follows. Recall that the distance from v_i to v_j, denoted d_{ij}, is the length of a shortest path from v_i to v_j. If there is no path from v_i to v_j, then d_{ij} is considered to be infinite, symbolized by ∞. The *distance matrix* of a digraph D, denoted[3] $N(D)$, or of a graph G, denoted $N(G)$, is the square matrix of order p whose entries are the distances d_{ij}. The next theorem gives the entries of this matrix.

Theorem 5.6. Let $N = [d_{ij}]$ be the distance matrix of a given digraph D or graph G. Then,

[3]We cannot use the first letter of the word "distance" for this matrix, as D already means "digraph," so N is used for "nearness." In general, one must avoid the two sins of notation, which are (a) using the same letter or other symbol with more than one meaning, and (b) using two or more different letters with the same meaning. Each of these two sins is worse than the other.

Structural models in anthropology

(1) Every diagonal entry $d_{ii} = 0$.

(2) $d_{ij} = \infty$ if $r_{ij} = 0$, and

(3) Otherwise, d_{ij} is the smallest power n to which A must be raised so that $a_{ij}^{(n)} \geq 0$; that is, so that the i,j entry of $A^n\#$ is 1.

The procedure for constructing the distance matrix N from the adjacency matrix A is illustrated for the digraph in Fig. 5.11. First of all, enter 0s on the diagonal $N(D)$, showing $d_{ii} = 0$. Next, enter 1 in $N(D)$ whenever $a_{ij} = 1$, thus showing the distance $d_{ij} = 1$. Taking higher powers of A, whenever $a_{ij}^{(n)}\# = 1$ and there is no prior i,j entry in $N(D)$, enter an n to show where $d_{ij} = n$. Finally, note that in $A^4\#$, every 1 occurs where there is already an entry in $N(D)$. Hence we enter ∞ in all remaining open locations, indicating all ordered pairs (v_i, v_j) for which there is no path from v_i to v_j in D.

The entries of the distance matrix provide the information needed to

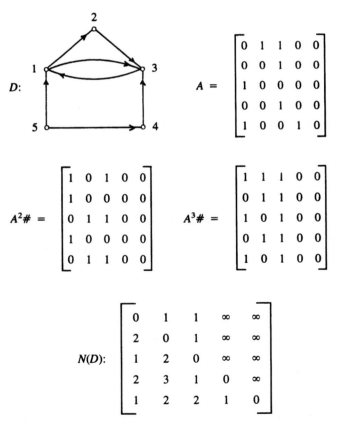

Fig. 5.11. The construction of the distance matrix.

Graphs and matrices

define all three measures of centrality in a connected graph G of a social network, as defined in Chapter 2. Consider, for example, the distance matrix of the graph of the Mayan ceremonial complex defined in Chapter 2.

$$
N(G) = \begin{array}{c}
 \\ \text{VI} \\ \text{VII} \\ \text{X} \\ \text{IX} \\ \text{VIII} \\ \text{XIV} \\ \text{XI} \\ \text{XII} \\ \text{V} \\ \text{IV} \\ \text{III} \\ \text{II} \\ \text{XVIII} \\ \text{XVII} \\ \text{XVI} \\ \text{XV}
\end{array}
\begin{array}{cccccccccccccccc}
\text{VI} & \text{VII} & \text{X} & \text{IX} & \text{VIII} & \text{XIV} & \text{XI} & \text{XII} & \text{V} & \text{IV} & \text{III} & \text{II} & \text{XVIII} & \text{XVII} & \text{XVI} & \text{XV} \\
0 & 3 & 5 & 6 & 1 & 2 & 4 & 5 & 2 & 3 & 4 & 5 & 4 & 3 & 3 & 4 \\
3 & 0 & 4 & 5 & 2 & 1 & 3 & 4 & 3 & 2 & 3 & 4 & 5 & 4 & 4 & 3 \\
5 & 4 & 0 & 3 & 4 & 3 & 1 & 2 & 3 & 2 & 3 & 4 & 5 & 4 & 4 & 3 \\
6 & 5 & 3 & 0 & 5 & 4 & 2 & 1 & 4 & 3 & 4 & 5 & 6 & 5 & 5 & 4 \\
1 & 2 & 4 & 5 & 0 & 1 & 3 & 4 & 1 & 2 & 3 & 4 & 3 & 2 & 2 & 3 \\
2 & 1 & 3 & 4 & 1 & 0 & 2 & 3 & 2 & 1 & 2 & 3 & 4 & 3 & 3 & 2 \\
4 & 3 & 1 & 2 & 3 & 2 & 0 & 1 & 2 & 1 & 2 & 3 & 4 & 3 & 3 & 2 \\
5 & 4 & 2 & 1 & 4 & 3 & 1 & 0 & 3 & 2 & 3 & 4 & 5 & 4 & 4 & 3 \\
2 & 3 & 3 & 4 & 1 & 2 & 2 & 3 & 0 & 1 & 2 & 3 & 2 & 1 & 1 & 2 \\
3 & 2 & 2 & 3 & 2 & 1 & 1 & 2 & 1 & 0 & 1 & 2 & 3 & 2 & 2 & 1 \\
4 & 3 & 3 & 4 & 3 & 2 & 2 & 3 & 2 & 1 & 0 & 1 & 4 & 3 & 3 & 2 \\
5 & 4 & 4 & 5 & 4 & 3 & 3 & 4 & 3 & 2 & 1 & 0 & 5 & 4 & 4 & 3 \\
4 & 5 & 5 & 6 & 3 & 4 & 4 & 5 & 2 & 3 & 4 & 5 & 0 & 1 & 2 & 3 \\
3 & 4 & 4 & 5 & 2 & 3 & 3 & 4 & 1 & 2 & 3 & 4 & 1 & 0 & 1 & 2 \\
3 & 4 & 4 & 5 & 2 & 3 & 3 & 4 & 1 & 2 & 3 & 4 & 2 & 1 & 0 & 1 \\
4 & 3 & 3 & 4 & 3 & 2 & 2 & 3 & 2 & 1 & 2 & 3 & 3 & 2 & 1 & 0
\end{array}
$$

The degree of a point is the sum of the 1s in its row of $N(G)$, and the eccentricity is the maximum entry in its row. As can be seen, plaza IV is most central in both senses. In such a graph the relative closeness of a point is the sum of all the entries in its row.[4] The betweenness definition of centrality, which refers to the frequency of occurrence of each point on the geodesics between all pairs of points, requires an enumeration of all the geodesics between each pair of points. The set of such geodesics, also called the *geodetic subgraph*, $D_g (v_i, v_j)$, can be found from the rows and columns of the distance matrix.

In the graph of the Orokaiva big man system analyzed in Chapter 2, it was shown that leaders ranked high on betweenness. To find the rank of each individual, it was necessary to find the geodetic subgraph of each pair of points in the graph of Fig. 2.12. We can define the procedure as follows, using point pair 1, 13 as an example. The next theorem, based in part on Flament (1963), is from Harary et al. (1965:139), but paraphrased for graphs instead of digraphs.

---

[4] The row sum of point v_i in the distance matrix of a digraph D (or a graph) has been called the "status" of this point (Harary 1959b) because it corresponds to this concept in organization theory. The column sum of v_i gives its "contrastatus," which can be regarded as the status of v_i in the organization obtained when its chart is turned upside down, that is, in the converse digraph D'.

Theorem 5.7. Let v_j be reachable from v_i in a graph H; that is, v_i and v_j are in the same connected component of H. The geodetic subgraph of H_g of v_i and v_j consists of all the points v_k, such that $d_{ik} + d_{kj} = d_{ij}$, and all the lines $v_r v_s$ of H, such that both $d_{ir} + 1 = d_{is}$ and $d_{rj} = d_{sj} + 1$.

The sums of the entries in the ith row of $N(G)$, which show d_{ik}, and the jth column, which show d_{kj}, disclose the points v_k in the geodetic subgraph D_g (v_i, v_j). To illustrate for D_g (1, 13), we take the first row and the 13th column of the distance matrix of the graph G in Fig. 2.11:

k	1	2	3	4	5	6	7	8	9	10	11	12	13	14	15	16	17	18	19	20	21	22
$d_{1,k}$	0	1	2	2	1	2	3	4	5	4	3	4	5	4	3	2	1	2	3	2	2	2
$d_{k,13}$	5	5	5	4	4	4	3	2	1	2	2	1	0	1	2	3	4	3	2	3	3	3
$d_{1,k}+d_{k,13}$	5	6	7	6	5	6	6	6	6	6	5	5	5	5	5	5	5	5	5	5	5	5

Since $d_{1,13} = 5$, all points except 2, 3, 4, 6, 7, 8, 9, 10 are in D_g (1, 13).

When the points v_k are arranged from left to right in order of their increasing distance from v_i and when a line is added from a point to its immediate right neighbor if the distance between them is 1, the result is the geodetic subgraph D_g (v_i, v_j). Fig. 5.12 shows D_g (1, 13) of the point pair 1, 13. These two points are on the left and right, and all other points lying on

$d_{1,k} =$ 0   1   2   3   4   5

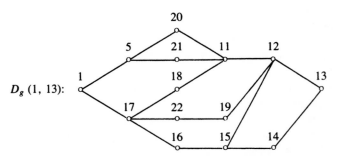

D_g (1, 13):

Fig. 5.12. The geodetic subgraph of point pair 1, 13, in the graph of Fig. 2.11.

1–13 geodesics appear between them in the order in which they occur in the graph G of Fig. 2.12.

In many situations, especially when a structure is large, a graph is best represented by its adjacency matrix. This matrix facilitates both the collection and the display of ethnographic data.[5] As an object of analysis, the adjacency matrix can be rearranged to produce patterns indicative of structural properties of a graph, such as bipartition and acyclicity, and it can be manipulated algebraically to yield higher-order matrices, such as reachability and distance. Those matrices, in turn, can be used to analyze other graphical properties, such as connectedness and centrality.

One of the five matrix operations defined was the transpose, that is, interchanging the rows and columns of a matrix A to get a new matrix A'. This operation, the converse, determines one type of structural duality in graphs.

[5]The adjacency matrix is also a basic tool in primate studies, where social structure is necessarily defined as "who does what to whom how often" (J. P. Hanby 1974).

6

Structural duality

But the nature of opposition is notoriously ambiguous.

David Maybury-Lewis, "Cultural categories of the Central Gê"

In *From honey to ashes,* Lévi-Strauss gives the following analysis of two South American myths on the origin of cultivated plants:

In one way, the Bororo and Machiguenga myths, instead of echoing each other, complete each other. According to the Bororo, man could speak to plants (by means of the whistled language) at a time when the latter were personal beings, capable of understanding such messages and growing spontaneously. Now, communication has been interrupted, or it is carried on through the medium of an agrarian god who speaks to man, and who is answered by man well or badly. The dialogue takes place, then, between the god and men, and plants are no more than the occasion of the dialogue. In the Machiguenga myths, the opposite is the case. Plants, being the daughters of the god and therefore personal beings, converse with their father. Men have no means of intercepting these messages: '*Los machiguengos no perciben esos eloros y regocijos*' (Garcia, p. 232); but since they are being talked about, they are the occasion of the exchange of messages. However the theoretical possibility of a direct dialogue existed in mythic times, when the comets had not yet made their appearance in the sky. But, at that time, plants were only half-persons, with the gift of speech but so indistinct in utterance that they were unable to use it for purposes of communication.

When completed one by another, the myths are seen, then, to form a total, multiaxial system. The Salesians point out that the Bororo whistled language has two main functions: to ensure communication between speakers who are too far away from each other to conduct a normal conversation; or to prevent eavesdropping by outsiders, who understand the Bororo language but are unacquainted with the secrets of the whistled speech.... This mode of communication is, therefore, both broader and more restricted. It is a super language for the actual speakers but an infra-language for outsiders.

The language spoken by plants has exactly opposite characteristics. When it is addressed directly to man, it is an incomprehensible muttering (M_{298}), whereas when it is used clearly, it by-passes man. He cannot hear it, although it is entirely concerned with him (M_{299}). The whistled language and indistinct speech therefore form a pair of contrasts. (Lévi-Strauss 1973:323-4)

Lévi-Strauss's diagrammatic rendering of the relations between these myths is shown in Fig. 6.1. These two diagrams are clearly digraphs in disguise. They are related to each other as complements: Informally, the second digraph has the same points as the first one, and it has arcs wherever the first one does not (and vice versa).[1] And they do indeed form a "total system" since a digraph D together with its complement $\bar{D}$ gives a complete symmetric digraph K_p. Complementation is one of three basic dual operations on graphs. Duality principles are fundamental in logic, mathematics, and physics, and judging by work such as *Mythologiques*, they are an integral part of structural anthropology.

Earlier, in a discussion of appropriate graph theoretic models of kinship

Bororo, M_{292}:

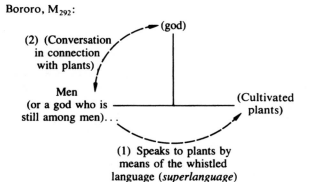

Machiguenga, M_{298}—M_{299}:

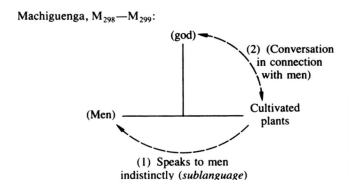

Fig. 6.1 Lévi-Strauss's (1973) diagram of the structural relations between two South American myths.

[1] The perpendicular lines are not part of the digraphs.

relations, we introduced the notion of structural duality to discriminate between heterogeneous types of relations in a bundle of relations. We now present an extended treatment. We begin by stating the properties of duality principles in general and then define three types of structural duality following Harary (1957): (1) existential, which is based on the operation of complementation; (2) directional, which is based on conversion; and (3) antithetical, which is based on negation. There are just two dualities derivable from these: uncontraduality, which combines the first and second, and anticontraduality, which combines the second and third operations. These distinctions serve to enrich and clarify the language of structural analysis, and they provide a source of transformation rules for determining structural relationships. We give an example of the use of such rules by analyzing two of Freud's origin myths. A variation of signed graphs known as marked graphs yields still further models for the analysis of symbolic transformations.

Duality

In general, any duality principle has two properties: (1) The dual of the dual of a statement is the original statement; and (2) the dual of a true statement is true. A classic example of duality occurs in set theory (Wilder 1952, Chapter 3), interchanging union and intersection, or equivalently in propositional logic (Wilder 1952, Chapter 9), interchanging disjunction and conjunction (the connectives "or" and "and"). Thus the dual of the logical DeMorgan Law:

$$\text{not } (p \text{ or } q) = (\text{not } p) \text{ and } (\text{not } q) \tag{1}$$

is

$$\text{not } (p \text{ and } q) = (\text{not } p) \text{ or } (\text{not } q). \tag{2}$$

Recall that the *union* of two sets A and B, written $A \cup B$, is that set whose elements lie in at least one of the sets A or B. Their *intersection*, $A \cap B$, is the set of all elements that are in both A and B. The *complement* $\bar{A}$ of A is the set of all elements under consideration that are not in A. Similarly, the dual of the commutative law for set addition:

$$A \cup B = B \cup A, \tag{3}$$

is the commutative law for set multiplication:

$$A \cap B = B \cap A. \tag{4}$$

A *self-dual* statement is one whose dual is the same statement. An example of a self-dual logical law is that of double negation:

$$\text{not } (\text{not } p) = p. \tag{5}$$

116

Structural duality

This law is self-dual, since there are no "and" and "or" connectives to interchange. Another self-dual statement, taken from set theory, is given by the equation:

$$A \cap (A \cup B) = A \cup (A \cap B). \tag{6}$$

This is sometimes called the "law of absorption," since both expressions are equal to the set A. On interchanging the operations union and intersection, each side of this equation is transformed into the other side.

In view of the DeMorgan laws, the interchange of the operations union and intersection can be applied by taking the complement or negation of both sides of a set theoretic or propositional equation, respectively. The law of double negation, which can be written for sets in the form:

$$\bar{\bar{A}} = A, \tag{7}$$

then assures us that this process is a dual one.

In Chapter 5 we introduced a matrix duality, the transpose A' of the adjacency matrix of a digraph. We can define three matrix dualities as follows. Let us denote the adjacency matrix of a digraph, a signed graph, or a graph by M. Then for directional duality $(M')' = M$, and

M' is the adjacency matrix of the converse of a digraph.

For antithetical duality, $-(-M) = M$, and

$-M$ is the adjacency matrix of the negation S^- of a signed graph S.

For existential duality, where J and I are defined as in Chapter 5,

$$(J - I) - [(J - I) - M] = (J - I) - (J - I) + M = M, \text{ and}$$

$(J - I) - M$ is the adjacency matrix of $\bar{G}$.

Structural duality

Existential duality is based on the operation of taking the complement of a graph or of a digraph. The *complement* of a graph G is that graph $\bar{G}$ having the same set of points as G, but in which two points are adjacent, or joined by a line, if and only if they are not adjacent in G. Thus the lines that are present in $\bar{G}$ are precisely those that are absent in G. To illustrate, we redraw Fig. 2.6, which shows all 11 graphs G with four points, as Fig. 6.2 to bring out the complementary pairs. In this new figure, the complement $\bar{G}_n$ of each graph G_n is G_{10-n}. Thus, for example, the complement of G_8 is $G_{10-8} = G_2$. The graph G_5 is self-complementary, since $\bar{G}_5 = G_{10-5} = G_5$. It is the only self-complementary graph with four points.

117

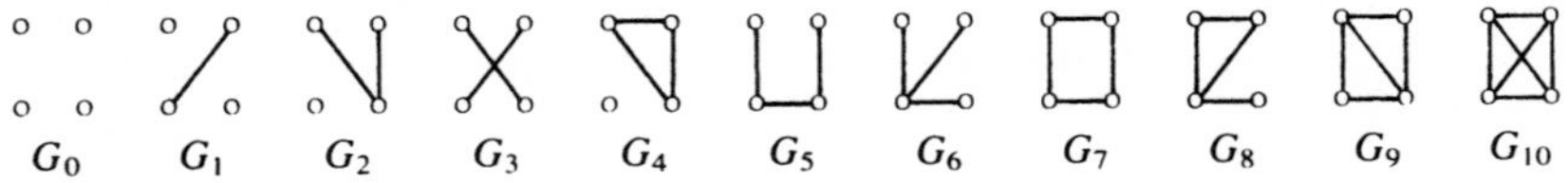

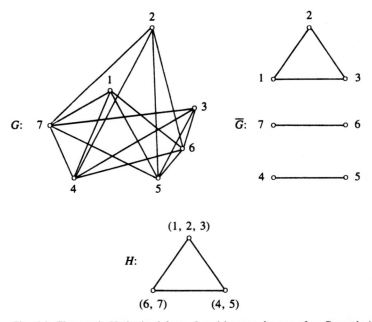

G_0 G_1 G_2 G_3 G_4 G_5 G_6 G_7 G_8 G_9 G_{10}

Fig. 6.2. The 11 graphs with four points drawn to illustrate complementarity.

An anthropological example is suggested by Crump's (1980) graphical depiction of social structure in a Chiapas village. In the graph G in Fig. 6.3, the points represent households and the lines represent marriage alliances between them. In the complement $\bar{G}$, the connected and complete subgraphs (cliques) represent exogamous units. The graph H is the reduced graph whose points represent the components of $\bar{G}$ and whose lines represent at least one marriage relation between them.

Fig. 6.3. The graph H obtained from G and its complement after Crump's (1980) model of marriage alliance in Chiapas.

We can use the graphs in Fig. 6.3 to extend our discussion of bipartite graphs and colorings of a graph. Bipartite graphs (also called bigraphs) have already been introduced. The *complete bipartite graph,* written $K(m, n)$ or $K_{m,n}$, has m points of one color and n points of another color, with two points adjacent if and only if they have different colors. Note that complete bigraphs with $p \geq 3$ points are not complete graphs. The best-

118

known complete bigraph is $K_{3,3}$ of Fig. 6.4, in which the points of the first color stand for three houses, h_1, h_2, h_3, and those of the second color for three utilities, u_1, u_2, u_3, standing for electricity, gas, and water. Of course, each house must be joined to all three utilities.

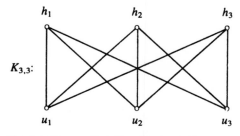

Fig. 6.4. The complete bipartite graph $K_{3,3}$.

The complete tripartite graph $K(m, n, r)$ is defined similarly (Harary 1969:23). Thus the graph of Fig. 6.3 is precisely $G = K(3, 2, 2)$, and its complement is $\bar{G} = K_3 \cup K_2 \cup K_2$. The graph H is obtained from G by identifying to a single point the set of points of each color in the complete tripartite graph G.

We note with interest that Crump uses graphs to represent both social and communication structures largely because his informants analyzed them as configurations of points and lines.

The *complementary digraph* $\bar{D}$ of a digraph D is defined by the same presence-absence consideration as for graphs. The only difference is that the lines of a digraph are directed from one point to another. An anthropological example was given in Fig. 6.1, Lévi-Strauss's diagram of relations between two South American myths. As an illustration of self-complementarity, all four digraphs with three points and three arcs are self-complementary, as shown in Fig. 6.5. In particular, the structures of the Bororo and Machiguenga myths in Fig. 6.1, the third pair of digraphs in Fig. 6.5, are self-complementary. Another illustration of complementarity, to which we will return in the course of this exposition, is given in Fig. 6.6. Note that the union of G and $\bar{G}$ and D and $\bar{D}$ is the complete graph or digraph K_p.

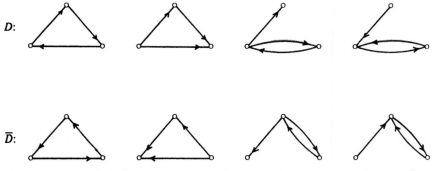

Fig. 6.5. The four digraphs D with three points and three arcs and their complements $\bar{D}$.

119

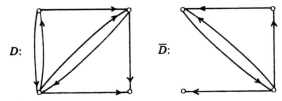

D: $\bar{D}$:

Fig. 6.6. A digraph and its complement.

Directional duality is based on the operation of taking the converse of a digraph. The *converse* D' of the digraph D is that digraph with the same points as D in which the arc uv occurs if and only if the arc vu is in D. The converse D' of the digraph D of Fig. 6.6 is given in Fig. 6.7. In addition Fig. 6.7 shows the converse of the digraph $\bar{D}$ of Fig. 6.6. We denote the converse of $\bar{D}$ by $(\bar{D})'$. Similarly, we write $\overline{D'}$ for the complement of the converse of D.

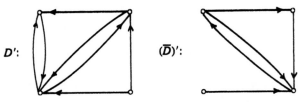

D': $(\bar{D})'$:

Fig. 6.7. The converses of the digraphs in Fig. 6.6.

In graph theory, a *unary operation* acts on a single graph G to obtain a well-defined graph. The same definition holds for a digraph D or a signed graph S. Thus each of the operations complement ($\bar{G}$ or $\bar{D}$), negation (S^-), and converse (D') and when S is a signed digraph (S') is unary. When an operation acts on two graphs, G_1 and G_2, to construct a uniquely defined graph G obtained from them, it is called a *binary operation*. Several such operations on graphs are presented in Harary (1969:22). One readily verifies the general theorem that the unary operations of converse and complement commute with each other, that is, $(\bar{D})' = \overline{D'}$. An anthropological example of directional duality was given in Chapter 1 in the digraphs of Fig. 1.5 depicting inverse hierarchical relations between characters in two different myths. Another example is provided by Peirce, as when he said: "Every relative has also a *converse*, produced by reversing the order of the members of the pair. Thus the converse of 'lover' is 'loved'" (Vol. 3, 330).

Antithetical duality is based on the operation of negation of a signed graph or digraph. The *negation* S^- of the signed graph S is obtained from S by changing the sign of each line or arc of S. Clearly, the negation of the negation of S is S itself. An example of antithetical duality is given in Fig.

Structural duality

6.8. Note that S is balanced, but S^- is not. An anthropological example is given in Fig. 3.8 by the diagonally opposite atoms of kinship.

There are two derived dualities. The combination of the complement and the converse results in uncontraduality, as illustrated by $\overline{D'}$ in Fig. 6.7. The combination of the negation and the converse yields anticontraduality, as exemplified in Fig. 6.9. This signed digraph represents a ritual of rebellion (Gluckman 1960) in which a normal relation S of dominance and reserve between men (M) and women (W) is temporarily replaced by the relation $S^{-\prime}$ of reverse dominance and license.

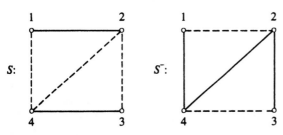

Fig. 6.8. Antithetical duality.

Fig. 6.9. Anticontraduality in ritual.

There are undoubtedly exact, multiple, and interesting senses in which, as Leach (1961) observed in "Time and false noses," ritual is "normal social life ... played in reverse."[2] Indeed, the theory of structural duality may serve to establish various of the logical possibilities of ritual transformation. Rodney Needham (1979) distinguishes three elementary forms of the manipulation and dramatization of a system of symbolic classification: inversion, disruption, and nullification. A correspondence between Needham's forms and our dualities is suggested by the following examples.

Inversion ↔ anticontraduality
A calendrical inversion of statuses can be traced from the saturnalia of Rome, when the master of a household might obey the orders of his own domestic slaves, down to the custom in the British army that on Christmas Day officers wait on their men at table and cheerily endure ribald insolences from them. (Needham 1979:40)

Disruption ↔ antithetical duality
Among Australian Aborigines there used to be set occasions on which their strict laws against incest were deliberately broken, as when men of the same moiety might

[2]There are also senses in which ritual B is ritual A played in reverse (Hage 1981).

exchange their wives. At one ceremony, sexual license among persons normally prohibited to one another was said to establish a general goodwill and to bring different groups closer together. (Needham 1979:42)

Nullification ↔ existential duality
At a certain point in the royal ritual of the Swazi, the king is ceremonially separated from his people; he performs certain mystically dangerous acts by which he "assumes the filth of the nation," thus intensifying his isolation. Then, at dawn on a particular day, he is washed and he emerges nude in front of the assembled people. The queen mother speaks of him as going alone through them; the queens say, "There is no other man who could walk naked in front of everybody." Beidelman, however, makes the significant point: "It is not nakedness denoting humiliation but a unique, lonely, denuded status outside any single social category." (Needham 1979:42)

Existential and antithetical duality are not combinable. If they were, then both the negation and the complement of a signed graph would have to be uniquely defined. The negation of a signed graph is completely determined, since it is obtained from the given graph by changing the sign of every line. The complement of a signed graph is not uniquely determined, since there are two possible choices for the sign of each line in it. This is illustrated in Fig. 6.10 where both Fig. 6.10(b) and (c) (among others) are possible candidates for the complement of Fig. 6.10 (a).

The theory of structural duality illuminates the complex notion of "opposite" which is so basic to various structural approaches in the social sciences. Freud, for example, used the model implicitly to discriminate emotional states when he said:

Loving admits of not merely one, but of three antitheses. First, there is the antithesis of loving–hating; secondly, there is loving–being loved; and in addition to these, loving and hating together are the opposite of the condition of neutrality or indifference. (Freud 1925:76)

These antitheses correspond to antithetical, directional, and existential dualities. In Chapter 3 the same concepts served to discriminate types of social relations for the purpose of making legitimate structural contrasts.

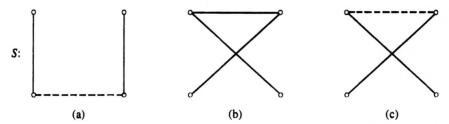

S:

(a)   (b)   (c)

Fig. 6.10. A signed graph and two possible candidates for its nonexistent complement.

Structural duality

We now show how structural duality can be applied to the analysis of transformations in myth. Our example is Freud's origin myths, the ontogenetic and phylogenetic versions of the Oedipus complex, treated as yet another variant of the Oedipus myth as analyzed along the lines proposed by Lévi-Strauss (1963d).

The irresistible motivation for such an analysis of Freud's myths is the self-characterization with which he concluded his most bizarre book, *Beyond the pleasure principle*: "What we cannot reach flying we must reach limping. The book tells us it is no sin to limp" (Freud 1961). Surely it cannot be accident that Freud chose this particular metaphor to conclude a book about the origin of the sexes from an original state of androgyny (the myth in Plato's *Symposium*);[3] clearly we have to do with a Viennese autochthonous hero, whose affliction signifies the same contradiction hypothesized by Lévi-Strauss for certain heroes in the Greek myths.[4]

According to Lévi-Strauss, the "purpose of myth is to provide a logical model capable of overcoming a contradiction (an impossible achievement if, as it happens, the contradiction is real)" (Lévi-Strauss 1963d:229). The contradiction is expressed by a structure of correlated oppositions that state the problem in direct and derivative form. The multiple representation of the same oppositions within and across different versions functions to "render the structure of the myth apparent." "Thus a myth exhibits a 'slated' structure, which comes to the surface, so to speak, through the process of repetition" (Lévi-Strauss 1963d:229).

Lévi-Strauss's theory can be characterized in part by saying that structural redundancy in myths is the converse of condensation in dreams: In myths, one latent thought gives rise to many manifest thoughts, whereas in dreams, many latent thoughts give rise to one manifest thought. Thus in the method of dream analysis, a "meagre, paltry and laconic" corpus (Freud 1913) is expanded by a process of contiguity (free association) to reveal multiple meanings, whereas in myth analysis a rich corpus is reduced by the discovery of similarity – "bundles of relations" – to reveal a single meaning expressed by a simple equation of the form $a:b::c:d$.

[3] In order to show that Eros, like the death instinct, is regressive in character, Freud appealed to the myth of original androgyny in Plato's *Symposium*. " 'The original human nature was not like the present, but different. In the first place, the sexes were originally three in number, not two as they are now; there was man, woman and the union of the two ....' Everything about these primaeval men was double: they had four hands and four feet, two faces, two privy parts, and so on. Eventually Zeus decided to cut these men in two, 'like a sorb apple which is halved for pickling.' After the division had been made, 'the two parts of man, each desiring his other half, came together, and threw their arms about one another eager to grow into one.'
"Shall we follow the hint given us by the poet-philosopher, and venture upon the hypothesis that living substance at the time of its coming to life was torn apart into small particles, which have ever since endeavoured to reunite through the sexual instinct?" (Freud 1961:51–2).

[4] A more extensive analysis, which includes Freud's successive theories of the "instincts," is given in Hage (1979c).

123

Table 6.1. *Components of the Oedipus myth according to Lévi-Strauss*

I	II	III	IV
Cadmos seeks his sister Europa, ravished by Zeus			
		Cadmos kills the dragon	
	The Spartoi kill one another		
			Labdacos (Laios' father) = lame (?)
	Oedipus kills his father, Laios		Laios (Oedipus' father) = left-sided (?)
		Oedipus kills the Sphinx	
			Oedipus = swollen-foot (?)
Oedipus marries his mother, Jocasta			
	Eteocles kills his brother, Polynices		
Antigone buries her brother, Polynices, despite prohibition			

Lévi-Strauss depicts the constituent units or bundles of relations in the Oedipus myth in the form of Table 6.1. In his analysis, the first column or bundle expresses the "overrating of blood relations," the second the "underrating of blood relations," the third the "denial of the autochthonous origin of man" (symbolized by killing chthonian beings), and the fourth the "persistence of the autochthonous origin of man." The rationale for the interpretation of the fourth column is provided by the "remarkable connotation of the surnames in Oedipus' father-line." They all hypothetically refer to "difficulties in walking straight and standing upright," which is significant because "in mythology it is a universal characteristic of men born from the earth that at the moment they emerge from the depth they either cannot walk or they walk clumsily" (Lévi-Strauss 1963d:215).

The contradiction the myth expresses is between the empirical knowledge of the facts of human procreation (= born from two) and a cultural and social theory (= born from one; that is, the belief in the autochthonous origin of humankind and a rule of unilineal descent):

Structural duality

Although the problem obviously cannot be solved, the Oedipus myth provides a kind of logical tool which relates the original problem – born from one or born from two? – to the derivative problem: born from different or born from same? By a correlation of this type, the overrating of blood relations is to the underrating of blood relations as the attempt to escape autochthony is to the impossibility to succeed in it. Although experience contradicts theory, social life validates cosmology by its similarity of structure. Hence cosmology is true. (Lévi-Strauss 1963d:216)

A basic principle of structural analysis is that there are not "true" but simply different versions that collectively define a myth. Freud's version is defined as follows:

Although the Freudian problem has ceased to be that of autochthony *versus* bisexual reproduction, it is still the problem of understanding how *one* can be born from *two*: How is it that we do not have only one procreator but a mother plus a father? Therefore not only Sophocles but Freud himself, should be included among the recorded versions of the Oedipus myth on a par with earlier or seemingly more "authentic" versions. (Lévi-Strauss 1963d:217)

Since Lévi-Strauss does not pursue this interpretation, three questions arise. (1) In what sense was the duality of the sexes a contradiction for Freud? (It was certainly a problem, for as he said, "Throughout history people have knocked their heads against the riddle of femininity" (Freud 1964:113). (2) What are the constituent units in Freud's myths that express this contradiction; in particular, what are the equivalents of the chthonian beings that must be overcome in order that man may live? (3) Since Freud had more than one version of the Oedipus myth, how are these related to each other?

For Freud, the contradiction clearly derives from the intellectual component of the Oedipus complex, the cognitive trauma produced by the "little investigator's" suspicions of heterosexual procreation that come up against his theories of sexual homogeneity.

But when the child [through the excitation in his penis] seems thus in a fair way to arrive at the existence of the vagina, and to attribute to the father's penis an act of incursion into the mother which should create the baby in the body of the mother, the inquiry breaks off helplessly; for at this point there stands in the way the theory that the mother possesses a penis like a man, and the existence of the cavity which receives the penis remains undiscovered to the child. One can readily surmise that the lack of success of this effort of thought facilitates a rejection and forgetting of it. *These speculations and doubts, however, become the prototype of all later thought-work on problems, and the first failure has a crippling effect for ever after* [emphasis added]. (Freud 1959:68)

Thus "all later thought-work" is subservient to the first principle in mental functioning, repetition compulsion (Freud 1961), in which an original traumatic situation is symbolically and repeatedly reinstated in an effort, often hopeless, to master it. Seen in this light, both of Freud's origin myths – the

125

Table 6.2. *Components in Freud's Oedipus myths*

I	II	III	IV
Sexual wish for opposite-sex parent	Death wish for same-sex parent		
Affectionate (seductive) wish for same-sex parent	Hostile, jealous attitude to opposite-sex parent	Consolidation of heterosexual identity	Castration anxiety
			Penis envy
	Sibling rivalry		
. . . .	. . . .	. . . .	. . . .
Father-daughter incest	Expulsion of the sons		
	Patricide		
		Exogamy	Totemism
			Matriliny
Overrating of blood relations	: Underrating of blood relations	:: Assertion of heterosexual distinctions	: Failure to maintain heterosexual distinctions

theory of the Oedipus complex, which explains the origin of the psychoneuroses and more generally of personality, and *Totem and taboo*, which explains the origin of culture – appear to be successive expressions of the problem of one sex or two, symbolized on the one hand by the over- and underrating of kinship and on the other by the assertion of sexual heterogeneity and the corresponding failure to maintain such a distinction.

Table 6.2 shows the constituent units of each myth, with the ontogenetic version above and the phylogenetic version below the dotted line. The first entries in columns I and II describe the Oedipus complex in its "simple positive form" (Freud 1913), and the second entries describe it in its "complete" form, which is more common (Freud 1960). In the Oedipus myth the equivalent of the chthonian monster is "innate human bisexuality" (Freud 1960), which is overcome by the assertion of psychological heterosexual identity in order that the individual may flourish. These distinctions are, however, symbolically denied – by fear of loss in the case of the boy, and desire for gain in the case of the girl. Both conditions persist in some form in later adult life – as conscience and as the woman's desire for the "penis-baby."

In *Totem and taboo* (Freud 1950), the monster is the original state of incest that is replaced by the institution of exogamy, which asserts social heterosexual distinctions in order that the group may flourish. However, the principle of exogamy, which says that humans are born from two, is contradicted by its "mysterious and notorious correlate" (Freud 1950), totemism, and also by matriliny, which both say that humans are descended from one. Although mature empirical knowledge contradicts infantile theories, these theories are ultimately vindicated by scientific discovery; the ultimate psychological reality is a residue of bisexuality, and the initial historical reality is humankind's denial of heterosexual origin.

One of the innovations of structural analysis consists of showing that different but related myths or different versions of the same myth constitute a group in a double sense; first, because they are about the same thing – convey the same message – and second, because they are transformations of homologous features. The Oedipus myth and *Totem and taboo* appear to constitute such a group. Both are about the problem of heterosexual distinctions, and the second is a transformation of the first globally with regard to level, and structurally with regard to the direction of asymmetry. The Oedipus myth is fantasy at the level of the individual, and *Totem and taboo* is reality ("In the beginning was the deed," Freud 1950) at the level of the group; asymmetric structures in the first myth are ascending, whereas in the second they are descending.

Table 6.3 represents the structures in the Oedipus myth and *Totem and taboo* as signed digraphs. The analysis treats as one unit columns I and II, the correlation between the over- and underrating of kinship expressed by positive and negative wishes and acts within and between generations. Another unit consists of columns III and IV, the correlation between the assertion and denial of heterosexual distinctions expressed by identifications and attitudes. The symbols 0 and 1 designate junior and senior generations; mP and fP, and m and f designate male and female parents and male and female children, respectively. Each structure in Freud's second myth is a simple transformation of a homologous structure in his first myth, defined by the operations of conversion and negation applied separately or in combination. The digraphs D_1 and D_2 represent the positive and negative relations in columns I and II of the Oedipus myth. Their converses are the descending positive and negative relations in *Totem and taboo*, D_1' and D_2'. D_3 represents the situation of sibling rivalry in the Oedipus myth in which positive relations to the senior generation induce negative relations in the junior generation. Its negation is patricide in *Totem and taboo*, in which negative relations to the senior generation induce positive relations in the junior generation, D_3^-.

The digraphs D_4 and D_5 use two kinds of signed relations: unit relations, U, which refer to identifications with some object, and evaluative relations, L, which refer to attitudes toward some object (Heider 1958). The rationale for the assignment of negative values to U relations is the one given by

Table 6.3. *Digraphs of the components in Freud's Oedipus myths*

I	II	III	IV
Directional duality (D')	Directional duality (D')	Directional duality (D')	Anticontra duality ($D'-$)
	+		Directional + antithetical combined
	Antithetical duality ($D-$)		

D_1: 0 ●———●◦ 1 D_2: 0 ◦— — —●◦ 1

D_4:
mP fP
(U) (L) $\times$ (L) (U)
m f

D_5:
mP fP
(U) (L) $\times$ (U) (U)
m f

D_3:
1
0   0

- -

D_1': 0 ●———◦ 1 D_2': 0 ●◦— — —◦ 1

D_4':
mP fP
m f

D_5':
mP fP
m f

D_3^-:
1
0   0

Heider; that is, the signification of disjunction as opposed to the absence of a relation. The dual operations consider only the fact of a positive or negative relation between the same set of points.

D_4 represents psychological heterosexual identity in the Oedipus myth, which consists of a positive identification with the same-sex parent and a positive (sublimated) attitude toward the opposite-sex parent. Its converse is exogamy in *Totem and taboo*, which consists of a positive identification of ancestry from each parent to each child, D_4'. D_5 combines castration anxiety and penis envy (the psychological male and female principles) in the Oedipus myth in which the son retains his positive identification with his father and rejects his positive (erotic) attitude to his mother whereas the daughter rejects her positive identification with her mother in favor of one with her father. The converse plus the negation of D_5 is D'_5^- which combines totemism and matriliny (the social male and female principles), in

Structural duality

which the mother has a positive relation, identification by ancestry, with each child, whereas the father has a negative (disjunctive) relation.

It appears that the novelty in Freud's second theory has the general characteristic of mythological thought – progression by the transformation or permutation of structure. The analysis of the relations between the relations agrees with the general rule according to which "each myth taken separately exists as the limited application of a pattern, which is gradually revealed by the relations of reciprocal intelligibility discerned between several myths" (Lévi-Strauss 1975:13).

Marked graphs

Another type of graph that has been implicitly used in the analysis of transformations in myths and also in dreams is a marked graph. In a *marked graph*, *M*, the points are designated positive or negative, in contrast to a signed graph, in which the lines carry these signs (Beineke and Harary 1978a). *Marked digraphs* (Beineke and Harary 1978b) are, of course, defined similarly. Here, changing the sign of each point is antithetical duality; changing the direction of each arc is directional duality; and doing both is anticontraduality. Interpreted empirically, the signed points can represent positive and negative events, actors, and so on. These operations are illustrated in Fig. 6.11.[5]

In an essay on structuralist method, Michael Carroll (1977) offers one interpretation of a generic transformation rule developed by Leach (1970) in his essay on biblical mythology. Carroll's aim is to show that a common

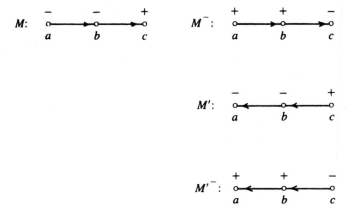

Fig. 6.11. A marked digraph and its duals.

[5]These and other operations on marked digraphs were introduced, and the resulting structures were enumerated, in Harary, Palmer, Robinson, and Schwenk, 1977.

structure underlies the stories in Genesis concerning the Fall of Man and Cain and Abel. In his analysis, the point of these stories is to establish a connection between the devaluation of kinship relations and the consequent promotion of exchange between groups on the one hand, and the overvaluation of kinship and the absence of exchange on the other. He thus interprets very generally Lévi-Strauss's (1969) theory, which derives culture from exogamy, by treating incest as an extreme case of the overvaluation of close kin relations. Leach's rule is this: "Substitute for each element its binary opposite" (1970:257). Carroll's interpretation is:

Carroll's Transformation Rule 2. Given a sequence of events, negate the outcome of each event and reverse the ordering of the events. (Carroll 1977: 675)

The significant events in the Cain and Abel myth are:

1. The devaluation of kinship (Cain kills his brother Abel).
2. Expulsion (Cain is forced to wander).
3. The acquisition of culture ("Cain acquires a wife and becomes associated with cities, the domestication of animals, blacksmithing and music").

Those in the story of Noah are:

1. The loss of culture (the flood).
2. Salvation (the arc).
3. Overvaluation of kinship (Noah seduced by his sons).

The structure of the first type of myth above is the graph M in Fig. 6.11 and that of the second, M'^{-}. This transformation rule is clearly an independent discovery of anticontraduality on a marked graph.

Subsequently, Adam Kuper (1979) used Carroll's rules in a proposed structuralist interpretation of dreams according to which the logic of dreams aims at the resolution of personal problems in the same way as the logic of myths aims at the resolution of cultural problems. Instead of treating a dream as a transformation of reality produced by the mechanisms of condensation, displacement, and secondary elaboration, Kuper views it as a dialectical process in which an argument reaches its conclusion through the successive transformations of an initial premise. Kuper's second rule is a variant of Carroll's:

Kuper's Transformation Rule 2. Given a sequence of events a, b, c, ... n:
a. produce a sequence $-a$, $-b$, $-c$, ..., $-n$; and/or
b. reverse the sequence. (Kuper 1979:650)[6]

---

[6]The fact that Kuper's sequence goes from a through n apparently does not imply that there are exactly 14 terms in it. The notation a_1, a_2, ..., a_n is a more convincing presentation of an n-term sequence.

Structural duality

This rule contains three dualities on marked graphs: antithetical, directional, and anticontra, as illustrated by M^-, M', and M'^- in Fig. 6.11.

It is interesting to find that both Carroll and Kuper regard their rules as specific instances of Lévi-Strauss's (1963d) generic formula, which is said to define the structure of a myth, regarded as composed of the set of its variants. According to Lévi-Strauss:

When we have succeeded in organizing a whole series of variants into a kind of permutation group, we are in a position to formulate the law of that group. Although it is not possible at the present stage to come closer than an approximate formulation which will certainly need to be refined in the future, it seems that every myth (considered as the aggregate of all its variants) corresponds to a formula of the type

$$F_x(a) : F_y(b) \simeq F_x(b) : F_{a-1}(y) \ . \tag{8}$$

(Lévi-Strauss 1963d:228)

Unfortunately, this formula has never been explained. Perhaps, then, rather than taking it literally, we should regard it as an evocative expression that can evidently conduce to, but is not equivalent to, any mathematical concepts. Lacking precise definition, it remains a tool of rhetoric, not of analysis.

Structural duality in graphs provides an explicit basis for discriminating logically different types of relations that underlie social and cultural structures, and it is a source of rules for relating apparently disparate structures as transformations of each other. A second basic model for analyzing structural transformations – a group – is defined in Chapter 8.

Graphs, signed graphs, and digraphs are all binary relations. In many situations, relationships between individuals, groups, events, beliefs, and so on are not all or none; rather, they differ in number, strength, probability of occurrence, and so on. Graphically, the lines have values, which gives a network.

131

7

Networks

Much that appears under the banner of network analysis fails to make use of its specific potentialities; we should be more abstemious in our use of the term.

John A. Barnes, *Social networks*

The question invariably arises as to whether graphs can accommodate social relations that differ in strength, number, or type. By suitably modifying a graph or digraph to permit the assignment of values to its lines or arcs, the answer is in the affirmative for many, although obviously not for all, kinds of empirical structures. Fig. 7.1, for example, depicts a social network, adapted by Patrick Doreian (1974), from Bruce Kapferer's (1969) analysis of a conflict in a Zambian work unit. The points represent men of the work unit, and the lines, social relations between them. The relations include (1) conversational exchange, (2) joking exchange, (3) job assistance, and (4) personal service. Any single relation is designated by Kapferer as "uniplex," and any multiple relation as "multiplex," represented by the thin and thick lines, respectively. The possible values of the lines are 1 or 2. (The values could also be 1, 2, 3, 4 if we chose not to lump multiplex relations.) The interest of this diagram, to which we will return, centers on the strength, as inferred from the multiplexity, of the direct and indirect links mobilized in support of two disputants.

The assignment of values to the lines, together with the relaxation of the stricture on loops, results in a very general model, of which a social network is only one interpretation. The values can represent such things as flows, probabilities, sequences, costs, and strengths in a variety of networks. We will offer three interpretations. The first is a social network, the one depicted in Fig. 7.1, where the values represent the strength of social relationships. The second is a capacitated network, where the values represent the amount of flow over a system of channels. A social structural example is Zachary's (1977) application of the Ford-Fulkerson (1957) algorithm on network flow to the analysis of fission in a small group. The third interpretation is a markov chain, where the values represent transition probabilities between different states in a system. An ecological example is Thomas's (1972) com-

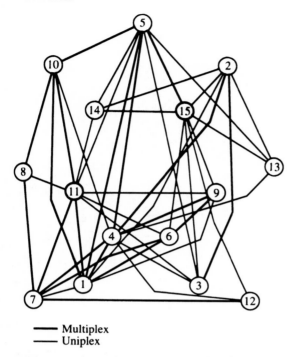

—— Multiplex
—— Uniplex

Fig. 7.1. A social network (from Doreian 1974; adapted from Kapferer 1969).

puter simulation of Great Basin Shoshonean subsistence patterns. All three exemplify the applicability of graph theory to processual analysis.

Networks and matrices

A *network N* consists of a relation on a finite set *V* of points, with its set of lines denoted as usual by *E*, but also including a "value" assigned to each line. If this value is always 1, we still have a relation. Thus relations form a special class of networks. The set *S* of values can be numerical or nonnumerical, as illustrated in Fig. 7.2 for two networks whose value sets are $S = \{1, 2, 3, 4\}$ and $S = \{a, b, c\}$. In certain situations, it may be convenient to use the set of positive integers or some subset thereof for nonnumerical values, for example, the designation of uniplex and multiplex social relations by 1 and 2.

To any network *N*, with points $v_1, v_2, \ldots, v_p$, we may associate a matrix *M* whose role is analogous to that of the adjacency matrix of a graph or digraph. The matrix $M = [m_{ij}]$ has $m_{ij} = 0$ if the line $v_i v_j$ does not appear in *N*; otherwise, m_{ij} is the value of the line $v_i v_j$. The matrix *M* is called the

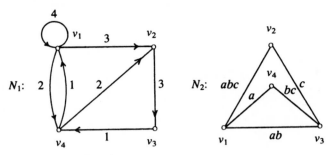

Fig. 7.2. Two networks.

value matrix of N. The diagonal entries m_{ii} need not, of course, be 0. The value matrix of N_1 in Fig. 7.2 is

$$
M = \begin{array}{c} \\ 1 \\ 2 \\ 3 \\ 4 \end{array}
\begin{array}{cccc}
1 & 2 & 3 & 4 \\
\end{array}
\left[
\begin{array}{cccc}
4 & 3 & 0 & 2 \\
0 & 0 & 3 & 0 \\
0 & 0 & 0 & 1 \\
1 & 2 & 0 & 0
\end{array}
\right]
$$

A *probability matrix* is a square matrix with nonnegative entries whose row sums are all 1, as illustrated in Fig. 7.3.

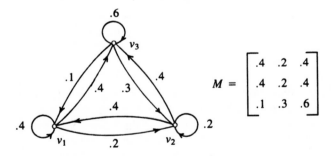

$$
M = \left[
\begin{array}{ccc}
.4 & .2 & .4 \\
.4 & .2 & .4 \\
.1 & .3 & .6
\end{array}
\right]
$$

Fig. 7.3. A network and its probability matrix.

Reachability in networks

The basic idea of a social network is that an individual's behavior – that is, his role or the performance of his role – may be affected by the ways in which he is directly or indirectly connected to other individuals and the ways in which they in turn are connected to each other. By using the analytical concept of a social network, as opposed to that of norm or institution, Kapferer was able to account for the differential success in the mobilization of

134

support in a dispute between two members of the Zambian group depicted in Fig. 7.1. His assumption was that "the amount of support a person achieves in a situation will be conditional on the structure and nature of his direct and indirect inter-personal relationships" (Kapferer 1969:209). Kapferer focused on a structure called a "reticulum," by which he meant "the direct links radiating from a particular Ego to other individuals in a situation, and the links which connect those individuals who are directly tied to Ego, and one another" (Kapferer 1969:182).

The dispute was between Abraham, point 11, and Donald, point 15, in the network in Fig. 7.1. It arose when the former accused the latter of rate-busting. It is clear from the figure that Abraham, who prevailed, has a stronger reticulum than Donald:

Abraham's ties in the situation cover individuals who are more closely interconnected with one another than those to whom Donald is tied, these interconnections tending to be multiplex in Abraham's reticulum, but uniplex, where they occur, in Donald's reticulum. (Kapferer 1969:232)

However, as Doreian (1974) emphasizes in his analysis of Kapferer's data, the mobilization of support may go beyond the 1-step links in a reticulum. He cites Kapferer himself, who says:

If the relationships of the persons to whom Abraham is directly connected by multiplex bonds are examined, it can be seen that they draw into their own reticulums, often by multiplex links, some of the individuals in Donald's reticulum who are not directly connected to Abraham. (Kapferer 1969:232)

The nature of Abraham's ties could be said to have had a cumulative effect which was to operate to Donald's disadvantage in the mobilization of support during the course of the dispute. Abraham was connected to individuals who in turn drew others to them, frequently by virtue of their multiplex links, and so drew away from Donald any support he might possibly have hoped for. (Kapferer 1969:234)

Thus it became important to examine reachability, connections at a distance greater than 1. In order to handle a large social network, Doreian proposes a modification of the reachability matrix as defined in Harary et al. (1965) and in Chapter 5 of this book.

Given a set of linearly ordered values such as the positive integers assigned to the lines of a network N, a "*path at level n* is simply a path where every valuation on a line in that path is greater than or equal to n" (Doreian 1974:250). In Fig. 7.4, path P is at level 2, and Q is at level 1.

To find the level of a path from v_i to v_j, the following matrix operations are defined.

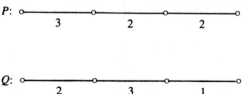

Fig. 7.4. Two paths at different levels.

Structural models in anthropology

Given two matrices of the same order, $A = [a_{ij}]$ and $B = [b_{ij}]$, let

$$A \vee B = [a_{ij} \vee b_{ij}] \quad \text{and} \tag{i}$$

$$A \wedge B = [a_{ij} \wedge b_{ij}] \tag{ii}$$

where $a_{ij} \vee b_{ij} = \max(a_{ij}, b_{ij})$, and $a_{ij} \wedge b_{ij} = \min(a_{ij}, b_{ij})$.

For an example of these matrices, consider

$$A = \begin{bmatrix} 1 & 2 & 3 \\ 3 & 1 & 2 \\ 2 & 1 & 3 \end{bmatrix} \quad B = \begin{bmatrix} 3 & 2 & 3 \\ 3 & 2 & 1 \\ 1 & 2 & 3 \end{bmatrix}$$

which gives

$$A \vee B = \begin{bmatrix} 3 & 2 & 3 \\ 3 & 2 & 2 \\ 2 & 2 & 3 \end{bmatrix} \quad \text{and} \quad A \wedge B = \begin{bmatrix} 1 & 2 & 3 \\ 3 & 1 & 1 \\ 1 & 1 & 3 \end{bmatrix}$$

(Doreian 1974:250).

The next matrix operation table defined, denoted $A * B$, uses both of these operations:

$$A * B = [\overset{n}{\underset{k=1}{\vee}} (a_{ik} \wedge b_{kj})]$$

The i, j entry of $A * B$ is $(a_{i1} \wedge b_{1j}) \vee (a_{i2} \wedge b_{2j}) \vee \cdots \vee (a_{in} \wedge b_{nj})$. In other words, "in determining the (i, j) element of $A * B$, each pair a_{ik} and b_{kj} (for successive values of k) has been examined, the smallest value of the pair selected, and then the largest of the selected values has been picked out" (Doreian 1974:251). For the two matrices A and B shown above,

$$A * B = \begin{bmatrix} 2 & 2 & 3 \\ 3 & 2 & 3 \\ 2 & 2 & 3 \end{bmatrix}$$

The following theorem gives the entries of a matrix M^n of a network N, or in Doreian's language, the entries of the adjacency matrix A of a valued graph G.

Theorem 7.1. Let G be a valued graph and A its adjacency matrix. Then $A * A * \cdots * A = A^{p} *$ gives the maximum level p step path between each pair of points in G (Doreian 1974:252).

This theorem is illustrated in Fig. 7.5, adapted from Doreian. Strictly

136

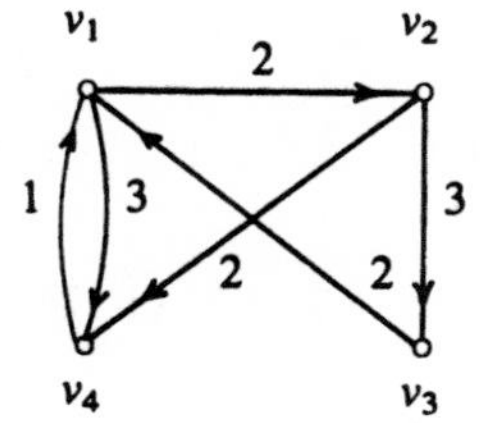

$$A = \begin{bmatrix} 0 & 2 & 0 & 3 \\ 0 & 0 & 3 & 2 \\ 2 & 0 & 0 & 0 \\ 1 & 0 & 0 & 0 \end{bmatrix}$$

$$\begin{bmatrix} 1 & 0 & 2 & 2 \\ 2 & 0 & 0 & 0 \\ 0 & 2 & 0 & 2 \\ 0 & 1 & 0 & 1 \end{bmatrix} \qquad \begin{bmatrix} 2 & 1 & 0 & 1 \\ 0 & 2 & 0 & 2 \\ 1 & 0 & 2 & 2 \\ 1 & 0 & 1 & 1 \end{bmatrix}$$

A^2_*:   2-step reachability A^3_*:   3-step reachability

$$\begin{bmatrix} 2 & 2 & 2 & 3 \\ 2 & 2 & 3 & 2 \\ 2 & 2 & 2 & 2 \\ 1 & 1 & 1 & 1 \end{bmatrix}$$

T: maximum reachability

Fig. 7.5. A network and its reachability matrices (from Doreian 1974).

(correctly) speaking, the word "path" in the preceding theorem is wrong [1] and must be replaced by "walk"; see Harary (1969:203). Fortunately, however, the next theorem is correct as stated because of the maximum symbol $\vee^n_{p=1}$, which enables paths to dominate walks.

Theorem 7.2. The matrix $T = \vee^n_{p=1} A^{p*}$ gives the highest reachability level between all pairs of points. (Doreian 1974:253)

This result is also illustrated in Fig. 7.5.

   Doreian applies these operations to Kapferer's network, shown in Fig. 7.1. As mentioned, uniplex links are simply contrasted with multiplex links so that the values of the lines are 1 or 2, with 0 for no line. The matrix A gives the 1-step links of each actor to every other one. Here Abraham is able to reach three individuals through multiplex links, and Donald, one. The matrix A^2 gives the 2-step links, and when A is combined with A^2, $A \vee A^2$, to

[1]As noted by E. R. Peay (1976).

give maximum reachability in one or two steps, the score is 8 to 4 in favor of Abraham, as can be seen from the matrix reproduced below. Going beyond the reticulum to consider 2-step reachability shows a more dramatic difference in the mobilization of support.

		1	2	3	4	5	6	7	8	9	10	11	12	13	14	15
Damian	1	2	2	1	2	2	2	2	2	2	2	2	1	1	1	2
Godfrey	2	2	2	2	2	2	1	2	0	2	1	1	1	1	1	1
Soft	3	1	2	2	2	1	1	1	1	1	1	1	1	1	1	1
Lotson	4	2	2	2	2	2	2	2	2	2	2	2	2	1	1	2
Jackson	5	2	2	1	2	2	2	2	2	2	2	2	1	1	1	2
Joshua	6	2	1	1	2	2	2	2	2	2	2	2	2	1	1	1
Andrew	7	2	2	1	2	2	2	2	2	2	2	2	2	1	1	1
Henry	8	2	0	1	2	2	2	2	2	1	2	2	2	0	1	0
Abel	9	2	2	1	2	2	2	2	1	2	1	1	1	1	1	1
Maxwell	10	2	1	1	2	2	2	2	2	1	2	2	1	1	1	2
Abraham	11	2	1	1	2	2	2	2	2	1	2	2	2	1	1	1
Stephen	12	1	1	1	2	1	2	2	2	1	1	2	2	1	1	1
Noah	13	1	1	1	1	1	1	1	0	1	1	1	1	1	1	1
Benson	14	1	1	1	1	1	1	1	1	1	1	1	1	1	1	1
Donald	15	2	1	1	2	2	1	1	0	1	2	1	1	1	1	2

Kapferer was concerned to show in his network analysis why the dispute, which signaled a deeper conflict, broke out between Abraham and Donald rather than between another pair, Benson and Abel, which would have been equally suitable given the external constraints of the work situation. Why did not Benson rather than Abraham make the accusation, and why was it not directed at Abel rather than Donald? One explanation lies in the differences in the mobilization of support using multiplex links, which are magnified when one considers both 1- and 2-step reachability, $A \lor A^2$, rather than just the reticula of the actors. At this level, as is shown in the matrix above, Benson has zero multiplex links, whereas Abraham has eight, Abel has six, and Donald has four.

This analysis assumes, of course, that individuals and paths of level n in a network are undifferentiated. We therefore regard it as a kind of first approximation similar to the model of mediation and power described in Chapters 2 and 5.[2] Analogous to our comment on the desirability of using graph theoretic models during fieldwork is Doreian's comment on identifying significant individuals in a network:

In analyzing the response of individuals to the activation of multiplex links, Kapferer remarked that Lotson, as the shop steward, is a powerful man and is, presumably,

[2] Doreian (1981) contains a further analysis of this network using the model of polyhedral dynamics.

an important person to have mobilized on one's behalf. This suggests that if it is possible through prior fieldwork to indicate those members of a network that would be crucial in this respect, then the procedures of the reachability matrices can be used to determine when, if at all, these persons are mobilized on behalf of particular individuals. (Doreian 1974:256)

Before leaving this section, comment is required on an unfortunate terminological confusion. In some contexts Doreian, even though his model is explicitly graph theoretic, uses the words "connectivity" and "reachability" as synonyms. They are not the same thing. The connectivity of a graph is an important and distinct structural concept, which bears on considerations here and on those in Chapters 2 and 4 as well.

The *connectivity* $\kappa = \kappa(G)$ of a graph G is the minimum number of points whose removal results in a disconnected or trivial graph. Thus the connectivity of a disconnected graph is 0, while the connectivity of a connected graph with a cutpoint is 1. The complete graph K_p cannot be disconnected by removing any number of points, but the trivial graph results after removing $p - 1$ points; therefore $\kappa(K_p) = p - 1$. Sometimes κ is called the *point connectivity*. Analogously, the *line connectivity* $\lambda = \lambda(G)$ of a graph G is the minimum number of lines whose removal results in a disconnected or trivial graph. Thus $\lambda(K_1) = 0$ and the line connectivity of a disconnected graph is 0, whereas that of a connected graph with a bridge is 1.

Flows in networks

Another anthropologically interesting way of interpreting a network is as a system of channels capable of carrying a flow of such things as messages, women, or goods, to use Lévi-Strauss's (1963a) communication metaphor of social structure. Each arc x of the network N represents a channel that can carry a flow from its first point f_x to its second point s_x. The *capacity of an arc* is the largest number of units that can flow through it during a given time period.

A network must have a source from which the units originate and a sink where they terminate. It is convenient to model the *source* by a transmitter t and the *sink* by a receiver r. It is also customary to have directed networks that are not only irreflexive (no loop arcs), but also asymmetric (no symmetric pairs of arcs). The reason is, of course, that in a flow network things need to be sent to some necessary destination, not from a place to itself or back and forth betweeen two places. Fig. 7.6(a) shows such a network, with the capacity indicated on each arc. In this example, every capacity is a positive integer.

A *flow f* in N from t to r results from assigning some numerical value to each arc. The numerical value as assigned must satisfy two conditions: (a) The flow value of each arc does not exceed its capacity (but may be zero);

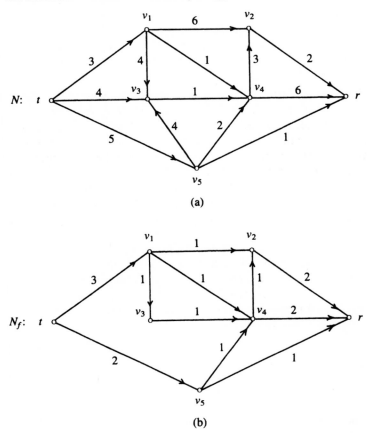

Fig. 7.6. A network N and a flow network N_f.

(b) for each point other than t and r, the flow indegree equals the flow out-degree. The *flow network* so obtained from N is denoted by N_f and is illustrated in Fig. 7.6(b).

In the network of Fig. 7.6(a), the outcapacity of the transmitter is 12, and the incapacity of the receiver is 9. Thus no more than 9 units can reach r, even if 12 are sent from t. But this does not guarantee that 9 units can actually get to r, since the capacities and structure of the intermediate arcs may not be sufficiently great. By the definition of a flow, it follows that in N_f the outflow from t and the inflow to r are equal, to a number called the *size of the flow*. Thus, the size of the flow f in network N_f of Fig. 7.6(b) is 5. But 5 is not the greatest size of all possible flows in N. A *maximum flow* in N from t to r is a flow of greatest size, that is, the largest number of units that can reach r from t within N, subject to the constraints imposed by the capacities of the arcs.

140

We next state the celebrated Ford-Fulkerson (1957) Maximum Flow Theorem (MFT), which determines the maximum flow in a network from t to r. But first we require some definitions.

If S is a set of arcs of N such that every path from t to r contains an arc of S, then S is a $t - r$ *cut set.* Thus in $N - S$ there is no (directed) path from t to r, whence the terminology "cut set." In Fig. 7.6(a), the arcs from t constitute a $(t - r)$ cut set, as do the arcs to r. The *capacity of a set* of arcs of N is the sum of their individual capacities. Thus the capacity of the cut set of arcs from t is 12. The *minimum cut capacity* from t to r is the smallest capacity among all $t - r$ cut sets.

Obviously no flow can exceed the minimum cut capacity from t to r, for the flow value in each arc cannot exceed its capacity. That there exists a flow from t to r whose size equals the capacity of some $t - r$ cut set is not obvious (sometimes called, in jest, the "harder half" of the proof) and is the content of the Maximum Flow Theorem, sometimes called the Max-flow Min-cut Theorem.

Theorem 7.3. In any network having a receiver r reachable from a transmitter t, the size of a maximum flow from t to r is equal to the minimum cut capacity from t to r.

This theorem can be explained by means of the network N in Fig. 7.6(a). Although the cut sets of arcs from t and to r have capacities 12 and 9, neither has the minimum value. For a careful search reveals the existence of a cut set with capacity 7; namely, the set of four arcs, $S = \{tv_1, v_3v_4, v_5v_4, v_5r\}$. It follows from the MFT that there can be no flow from t to r with a size greater than 7. Thus, when we can find a flow of 7, we know that its size is maximum and so the $t - r$ cut set S must have minimum capacity. The flow network N_f in Fig. 7.7 shows a $t - r$ flow in N of size 7. Note that for every other point v_i, the inflow and outflow are equal. The heavier arcs of this figure constitute a minimum cut set of N and have flow values equal to their capacities.

Finding the maximum flow by locating minimum cut sets can be done using the Ford-Fulkerson (1962) algorithm. This algorithm also determines whether a network has a unique minimum cut set. A little experimentation with the network in Fig. 7.6 shows that it does not have such a set: Instead of the arcs in the minimum cut set $S = \{tv_1, v_3v_4, v_5v_4, v_5r\}$, one could, for example, equally use $S_1 = \{v_1v_4, v_3v_4, v_2r, v_5v_4, v_5r\}$ and still get a maximum flow of 7. The determination of the existence of a unique minimum cut is important in the anthropological application of networks we now describe.

Wayne Zachary (1977) uses the model of network flow to study fission in a voluntary association. In so doing, he extends the study of fission beyond the usual confines of kinship and introduces into anthropology a new and

141

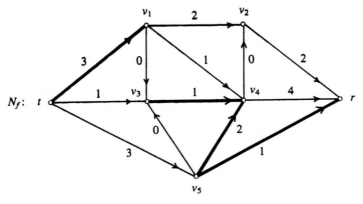

Fig. 7.7. Another flow network N_f.

general type of structural model. The group is a university karate club, consisting of 34 members, which split into two subgroups as a result of an internal dispute over club policy. The dispute was between the club president and the instructor and occurred when the latter tried to raise his fees and the former denied his authority to do so. This particular issue became a general ideological one with supporters on each side. The outcome was an irrevocable split in the club and the formation of a new one, led by the instructor and consisting of some members of the old club. Zachary applies the Maximum Flow Theorem to predict the locus of fission and the membership of the two resulting subgroups.

He conceives of the club as a communication network in which there are unequal flows of sentiment and political information among members. Each of the two leaders, the president and the instructor, is regarded as a transmitter in a flow of information concerning club activities, and each is in the role of receiver to the other. Because the flow of information is uneven, the group is regarded as most vulnerable where the flow is most constricted; that is, at the minimum cut. This is characterized as a kind of "bottleneck," and the prediction is made that the membership of an individual in the two new subgroups will be determined by which side of the cut he is on.

Zachary's specific procedure is as follows. He first constructs the graph G, shown in Fig. 7.8, of the communication network of the club. The points represent club members, and the (undirected) lines represent relations of extracurricular friendship interpreted as a channel for the flow of incidentally communicated information concerning club activities. The information included such things as dates of club meetings and notions about club ideology. He then assigns values to the lines representing the number of different contexts in which two members have occasion to communicate

142

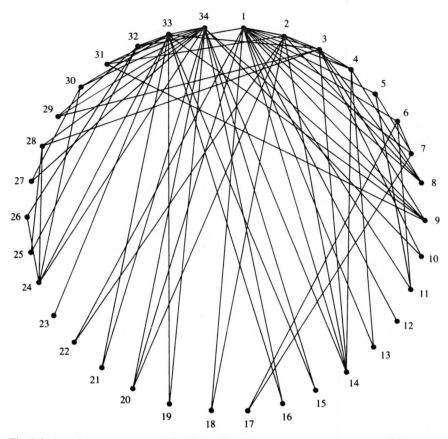

Fig. 7.8. A social network model of relationships in a karate club (from Zachary 1977).

such information. His assumption of linearity "claims that the likelihood of [an] item of [incidentally] communicated information being communicated to any individual increases linearly with the number of contexts in which interaction with that individual takes place" (Zachary 1966:460). There are eight different contexts, such as interaction in the same classes, bars, and karate tournaments. Each context in which u and v associate adds a value of 1 to their line capacity. Thus we can think of a flow of information across the lines of the network from a transmitter (leader) to a receiver (the other leader), with the greatest restriction of the flow at the minimum cut.

Since Zachary's matrix representation of this 34-point network must of necessity be analyzed by computer, it will not be reproduced here. Instead, we show in Fig. 7.9 a miniature version of his model. The heavy lines represent the minimum cut, with the members on the two different sides enclosed by triangles and circles. This minimum cut is unique; if it were not, there would be no basis for predicting subgroup membership.

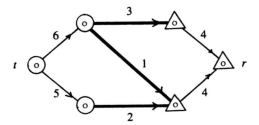

Fig. 7.9. A flow network with a unique minimum cut.

There is an interesting historical note on the Ford-Fulkerson Theorem. Menger (1927), while working on a difficult topological problem involving infinite sets of curves, inadvertently discovered the easier finite case that fully anticipated the Maximum Flow Theorem by 30 years. Very recently, Menger (1981) wrote a personal account of his research leading to this result, which appeared in König (1936) only because Menger visited Budapest to give a lecture and told König about it. Menger's Theorem has many equivalent forms, which have different settings and statements (Harary 1969, Chapter 5). We now give the line form of Menger's Theorem for undirected graphs.

Theorem 7.4. Let G be a connected graph, with t and r being two of its points. Then the maximum number of line-disjoint paths joining t and r equals the minimum number of lines whose removal separates t and r; that is, disconnects G in such a way that t and r are in different components.

This form of Menger's Theorem is easily modified to digraphs by taking t and r as points such that r is reachable from t and by changing lines to arcs and hence paths to directed paths, and "separates t and r" to "makes it impossible to reach r from t." To show convincingly how the line form of Menger's Theorem implies at once the MFT for networks with integer capacities requires the concept of multigraphs.

A *multigraph* differs from a graph only in that more than one line is permitted to join the same pair of points; the same holds for arcs in a *multidigraph*. For instance, Fig. 7.10 shows the multidigraph obtained from the cut set of arcs shown in Fig. 7.9 on replacing each of these arcs, x, and its capacity, c, by an assemblage of c different arcs from fx to sx. Fortunately, the c values occurring here are only 3, 2, and 1, so it was not necessary to draw an unrecognizably large number of arcs for this purpose. This

Fig. 7.10. The unique minimum cut set of Fig. 7.9 displayed as a multidigraph.

144

illustrates the usefulness of the network model: It is easier to write the number 20 on an arc x than to draw twenty arcs from f_x to s_x.

Now we can stipulate the precise way in which the MFT for integer capacity networks follows from this form of Menger's Theorem. This is based on the correspondence between multigraphs and the networks shown above. Referring to both Theorem 7.4 and Fig. 7.10, each of the line-disjoint paths from t to r in the multidigraph of network N contributes 1 to the flow value, and hence the maximum number of arc-disjoint directed paths from t to r equals the maximum flow. Fig. 7.10 already shows by illustration how the minimum cut capacity equals the minimum number of arcs whose removal from the multidigraph makes it impossible to go from t to r.

Markov chains

A *markov chain*, or more briefly a *chain*, is a network with directed loops permitted in which the value of each arc is a positive number, and the sum of the values of the arcs from each point is exactly 1. Thus the matrix of a chain, called its *transition matrix*, is a probability matrix. The points of a chain are called its *states*. The value of an arc x is the conditional probability that if the present state is fx, then the next state will be sx. The theory of markov chains studies sequences of events with a given distribution of initial probabilities that are used only to determine the initial state. Then a chain has the single property that the probability of the next event in a sequence depends only on the present event and not on the preceding ones.

In both anthropology and archeology, use is sometimes made of a systems approach to the analysis of culture; for example, Rappoport (1968), Clarke (1968). When explicit theoretical propositions concerning cultural systems or subsystems cannot be tested directly, they can often be tested by means of simulation. In Chapter 3 we described a simulation performed to test a theory of alliance formation using signed graphs, or more accurately signed networks, as a structural model. We now describe a simulation undertaken to test a theory of subsistence behavior using a network, more specifically a markov chain, as a model. We refer to David Thomas's (1972) study of Julian Steward's (1938, 1955, 1970) theory of Great Basin Shoshonean economic patterns.

Steward proposed that a specific type of seasonal round made life possible in the Great Basin, but also set a limit on the elaboration of Shoshonean social organization:

The main point [of Steward's theory] is that the cultural ecology of the Great Basin had a socially fragmenting effect upon the aboriginal population. The constraints imposed by this harsh, *unpredictable* environment were such that traditional institutions other than nuclear family were notably absent. The Shoshoneans depended according to Steward (1970, p. 116) upon a *multiple subsistence* pattern which exploited contiguous but dissimilar micro-environments. A well-defined seasonal round permitted the Shoshoneans to extract a living from the deserts of the Great Basin. (Thomas 1972:674)

Steward's theory has been generalized to interpret numerous other Amerindian groups, but as Thomas points out, it has never really been tested because the seasonal round Steward describes was not directly observed, but was instead constructed from informants' memories. Even granting the veridicality of this "memory culture," it may not have corresponded at all to the precontact situation.[3] Thomas therefore proposes a test through simulations of the various components of the Shoshonean subsistence pattern.

To determine the availability of the piñon nut, the staple food of the Shoshone, a markov chain is used. According to Steward, this crop is highly erratic:

Each tree yields but once in three or four years. In some years there is a good crop throughout the area, in some years virtually none. In other years, some localities yield nuts but others do not. When a good crop occurs, it is more abundant than the local population can harvest. (Steward 1938:27)

When the local crop failed families travelled to the nearest locality with a successful nut crop. Winter villages were located within a few hundred yards of the nut caches, because of the difficulty of transporting the bulky nuts. (Thomas 1972:684)

Although each local crop was unpredictable, the regional crop was, in theory, always sufficient, provided families were willing to travel up to 40 or 50 miles from their home base (winter village) to harvest it.

Thomas's assumption is that piñon nut production is markovian; that is that the relative abundance of the crop in one year depends in part on its abundance in the immediate preceding years, but not on any years before that. This relationship is attributed to the "long maturation period of the piñon ovulates, the mechanisms behind which are still a mystery to forest scientists." We note that a model such as this one could be modified in various ways, depending on particular circumstances and available information. If the crop were partially dependent on the penultimate as well as the ultimate preceding year, the model would be a second order markov chain, and if other factors, for example rainfall, were taken into consideration, the model would be an nth order chain (Feller 1950). In some situations, a simple markovian model may represent a kind of first approximation.

Fortunately, there are some data available from another but presumably comparable part of the Southwest on piñon production. To help local piñon businesses forecast labor demands, the U.S. Forest Service in Tucson kept records of piñon densities in a number of different districts over a period of

---

[3]One eminent Great Basin scholar has facetiously described ethnography in this area as, "How does your grandfather remember it?"

Table 7.1. *Piñon nut crops in the Southwest national forests from year S to year (S+1), 1940–7*

	A, Failure	B, Fair	C, Good	Total	Percentage
A, Failure	488	26	66	580	79
B, Fair	37	5	4	46	6
C, Good	93	8	9	110	15
Total				736	

Source: After D. H. Thomas (1972).

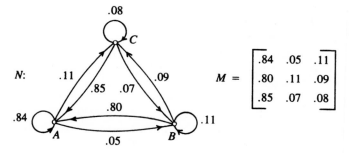

$$M = \begin{bmatrix} .84 & .05 & .11 \\ .80 & .11 & .09 \\ .85 & .07 & .08 \end{bmatrix}$$

Fig. 7.11. A markov chain model of piñon nut production (from Thomas 1972).

eight years. These observations are combined in Table 7.1 in a format enabling the construction of the chain, shown together with its transition matrix in Fig. 7.11.

The simulation of piñon nut productivity is performed by using the Monte Carlo technique of Ulam (1951). This method uses a random number generator to simulate actual data. The probabilities in the transition matrix are converted to random numbers by the following device. A 3×3 matrix is now written in which the 100 numbers from zero (written 00) through 99 appear in each row within one of the three intervals in that row. These intervals are of the form 00–*xx*, *xx*–*yy*, *yy*–99. Then a 100-numbered random device such as a fair roulette wheel is invoked and one of the numbers, say 86, comes up. One looks into the row corresponding to the present state (*A*, *B*, *C*). If it is *A*, one inspects the first row to locate the interval containing the number obtained randomly (here 86), thus determining that the next state is *B* because 86 occurs in the interval 84–88. This is interpreted to mean that if the system is in state *A* in year *S* and the random number 86 is drawn, it will be in state *B* in year *S*+1 (a failure followed by a fair crop).

$$\begin{bmatrix} 00\text{–}83 & 84\text{–}88 & 89\text{–}99 \\ 00\text{–}79 & 80\text{–}90 & 91\text{–}99 \\ 00\text{–}84 & 85\text{–}91 & 92\text{–}99 \end{bmatrix}$$

In order to start the run, an initial state must be chosen. This is accomplished by converting the overall ratio of failure, and good years in Table 7.1 (79, 6, and 15 percent, respectively), to intervals of random numbers: 00–78, 79–84, 85–99. A trial, then, consists of two steps: First, the selection of a random number to determine the initial state of the system (S), and then the selection of a random number to determine which state this converts to ($S + 1$). The results of a 200-year trial simulation show a good crop every 7.7 years and a fair (acceptable) crop every 5.4 years. This accords with Steward's general characterization of piñon nut production in the Great Basin.

The second step in the analysis is the simulation of regional production; that is, production in contiguous microenvironments. This requires the definition of a set of homogeneous areal units – "minimal piñon groves." For the Great Basin, Thomas estimates a size of 207 square miles, an area sufficient to span all vegetation zones.[4] Given the independence of production in each unit, the question then concerns the distance that must be traveled to obtain an adequate – good or fair – harvest every year.

Fig. 7.12 shows a map of contiguous piñon groves represented as an optimal hexagonal model.[5] The star depicts the home base, and the circles, the increasing radius of foraging. The relation between the number of piñon groves and the radius of foraging is exponential, as shown in Fig. 7.13(a). When piñon productivity is simulated in all available groves, the relation between the probability of success in production and the radius of foraging tends to the asymptotic value at 30 miles, as shown in Fig. 7.13(b). According to Thomas,

This agrees remarkably well with Steward's information that Shoshoneans preferred to remain within about 20 miles of their winter village (1938, p. 232) but that they were willing to travel up to 40 or even 50 miles, when necessary, to secure a good piñon harvest (1938, p. 101). The possibility always exists that there are no successful areas in the entire system, but this remains an unlikely event. (Thomas 1972:689)

---

[4]The estimated size depends on a correction factor: "The minimal homogeneous units (ranger subdistricts) are 62 square miles (37,200 acres). This figure represents a pure forest stand, the area of prime interest to the Forest Service. But in the Central Great Basin, topographic relief is so sharp that the piñon-juniper zone generally forms only a narrow belt on the mountain flanks.... It becomes necessary to correct the Southwestern average to a figure more suitable to central Nevada conditions. From a Soil Conservation Service vegetational map of the region, I estimate the piñon acreage to cover about 30% of the total land surface. The correction factor changes the minimal piñon grove from 62 to 207 square miles. This figure spans all vegetational zones in the Great Basin study area" (Thomas 1972:687).

[5]Fig. 7.12 is a Löschian optimal-hexagonal model, chosen because "(1) hexagons allow the greatest amount of packing of regular cells on a plane, (2) both total edge length and accessibility (distance from center to all faces) are minimized..." (Thomas 1972:688–9). Plato knew not only the five regular solids (tetrahedron, cube, octahedron, dodecahedron, icosahedron), but also that exactly three regular polygons can pave an infinite plane: triangles, squares, and hexagons.

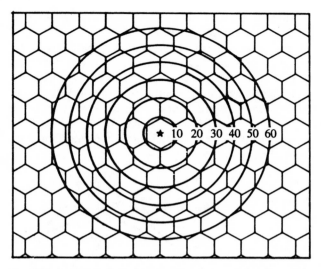

Fig. 7.12. A model of contiguous piñon groves and radius of foraging from home base (•) (from D. H. Thomas 1972).

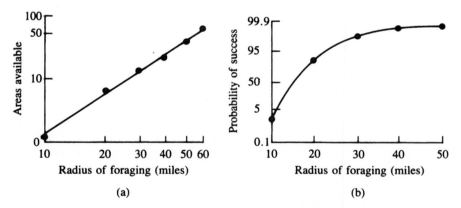

Fig. 7.13. (a) Radius of foraging plotted against areas available; (b) probability of success (from D. H. Thomas 1972).

Thomas's study is a very simple but nonetheless interesting and suggestive application of a markov chain. A general discussion of the model is contained in Chapter 14 of Harary et al. (1965) and in Kemeny and Snell (1960). As an illustration of the potentially broad application of such structural models, we note that although markov chains are most often used in the physical and life sciences, the first application was evidently in the human sciences. According to Olnick (1978), Andrei Andreevich Markov (1856–1922), after whom the model is named, published the first modern paper on mathematical linguistics when he applied his theory to the distribution of vowels and consonants in Pushkin's *Eugene Onegin.*

When a graph is modified so as to permit loops at its points, and the assignment of values to its lines, a general model called a network becomes available for the analysis of structures whose relationships are weighted. It thus becomes possible to analyze such things as multiplex relations in social networks, the flow of information in communication networks, and the direction and intensity of change in social systems. Again, the representational and matrix resources of graph theory assist in the comprehension and analysis of structural properties.

Fig. 7.14 from Harary (1966) shows a rather different kind of network than the ones studied so far. It depicts the dramatic structure of the play *A severed head* by Murdoch and Priestly, based on the novel by Iris Murdoch (1961). The six principal characters are:

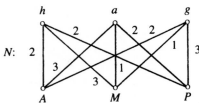

Fig. 7.14. The sexual consortium of *A severed head.*

Martin Lynch-Gibbon (a wine merchant)
Georgie Hands (his mistress)
Antonia (his wife)
Palmer Anderson (a psychoanalyst)
Alexander (Martin's brother, a sculptor)
Honor Klein (Palmer's half-sister, an anthropologist)

The points in this complete bigraph $K_{3,3}$ represent the characters, uppercase letters for male and lowercase for female, and the lines represent consort relations between them. The numbers on the lines represent the approximate state of the relation with the passage of time. For example, in the opening scene Martin has just consorted with his loving mistress Georgie and believes himself to be still on good terms with Antonia, his wife, thus accounting for the marks of 1 on the lines gM and aM. Since all possible heterosexual pairings occur in this consortium, we have a permutation of some kind. In Chapter 8 we define permutation groups, a basic underlying model in structuralism.

8

Graphs and groups

> Over and beyond the schemes of atomist association on the one hand and emergent totalities on the other, there is, however, a third, that of operational structuralism. It adopts from the start a relational perspective, according to which it is neither the elements nor a whole that comes about in a manner one knows not how, but the relations among elements that count.
>
> Jean Piaget, *Structuralism*

In structural analysis, algebraic metaphors are commonly used to convey the idea of sets of concepts or practices related by transformation rather than by simple difference or similarity. Words like "group," "permutation," "inverse," and "boolean" occur, and there may even be mathematical symbols like x, $-x$, $1/x$, $-1/x$. In anthropology, as contrasted, say, to genetic epistemology, group models are evoked rather than defined. But they nonetheless play an important role in the analysis of symbolism, as may be illustrated by two examples from *Mythologiques*.

In *The origin of table manners*, Lévi-Strauss analyzes beliefs about the outcomes of complementary male activities among a number of North American Indian tribes by using the analogy of the Klein group:

The possession of a scalp ensures success in war, while the ingestion of female scurf leads to lack of success in hunting. According to M_{493b}, the non-ingestion of the liver is the pre-condition of the husband's success in hunting (Hoffman, pp. 182–5; Skinner-Satterlee, pp. 399–400). Lastly, menstrual blood causes lack of success in fighting; the Plains Indians would remove the sacred bundles used for military rites from a tent when there was a menstruant woman. We thus arrive at a kind of Klein group, if we give the values x, $-x$, $1/x$, $-1/x$ to the scalp, the dandruff, the liver and the menstrual blood, respectively. (Lévi-Strauss 1968:403)

The group model is evoked a little later in the same work, where Lévi-Strauss shows an even more surprising association of menstrual blood, sperm, scalps, and honey, this time in connection with a system of affinal exchange. Referring to his previous demonstration of an analogy between menstrual blood and honey and his subsequent discovery of an association between honey and sperm, he says:

151

This extraordinary inversion of a system which I have shown to be characteristic of a vast territory stretching from Venezuela to Paraguay does not contradict my interpretation, but instead gives it an additional dimension, and allows the completion of the Klein group, of which, so far, the fourth term was missing. Sperm is *that which ought* to be transmitted from husband to wife; menstrual blood is that *which ought not* to be transmitted from wife to husband. As I showed in *From honey to ashes,* honey is something which has to be transmitted by the husband to the wife's relatives, and which, consequently, goes in the same direction as sperm, but further. Similarly, it has been established in the present volume ... that scalps also, are transmitted from husband to wife, and more often to the wife's relatives. We thus obtain a generalized four-term system, in which there is a correspondence between menstrual blood and sperm on the one hand, and between scalps and honey on the other. The husband transfers the sperm to his wife, and through the intermediary agency of his wife, he transfers the honey to his parents-in-law, as compensation for the wife he has received from them. Unless she is a sorceress ... the wife does not transfer menstrual blood to her husband. The husband, for his part, transfers the scalp to his wife's relatives in order to prevent the non-transference of menstrual blood taking on the significance of a non-transference of the woman herself by her parents, as if they were repudiating the agreement implicit in the marriage. (Lévi-Strauss 1968:412)

This passage on Amerindian mythology has been quoted at length, because in addition to the reference to groups, it contains an interesting model for the analysis of sex and affinity similar to the one we will provide in this chapter and similar also to our own group theoretic approach to the analysis of pollution beliefs in New Guinea (Hage and Harary 1981b).

A general reference to group models, and this time to boolean algebra as well, occurs in Lévi-Strauss's (1976b) essay on structuralism versus formalism *à propos* the work of Vladimir Propp (1968). Here he proposes that Propp's formal analysis, whereby tales are classified into so many distinct types and then generalized by progressive abstraction, should be abandoned in favor of a structural analysis that would treat tales and myths as a transformation group. The result would be not a list of types, but

a matrix with two or three dimensions or more:

w	$-x$	$1/y$	$1-z$	...
$-w$	$1/x$	$1-y$	z	...
$1/w$	$1-x$	y	$-z$	...
$1-w$	x	$-y$	$1/z$	...

and where the system of operations would be closer to boolean algebra. (Lévi-Strauss 1976b:137)

The symbols above have been characterized as "mathematically meaningless" (Regnier 1971), but mistakenly so, for their underlying significance is precisely the multiple oppositions shown in an *n*-cube, a graphic representation of a boolean group.

Lévi-Strauss's analyses of myth raise the following questions. What is a group? What is a Klein group? What is a boolean group? What is a permutation or transformation group? Explication of these concepts will reveal a productive interaction between graphs and groups. Anthropological illustrations include the culinary symbolism of gender relations in a New Guinea society and two divinatory systems, one of which exemplifies the kind of data that would actually be pertinent to evaluating Piaget's claim concerning the nonoperational basis of primitive thought, and the other of which shows a historically interesting isomorphism between primitive classification and modern mathematics.

Group models

Group theory is of fundamental importance in all fields of structural studies because groups define the systems of transformation that underlie the symmetry of empirical structures. In the natural sciences, group models are basic to crystallography (for the classification of crystalline structures), and are widely used in chemistry (for the study of molecular symmetry) and in physics (for the specification of elementary particles); indeed, in Martin Gardner's (1980) phrase, "wherever there is symmetry there is a group." In fact, all graphical enumeration and its applications to fields ranging from architecture to zoology is based on group theory (Harary and Palmer 1973). In the social sciences, group models figure most explicitly in Piaget's genetic epistemology. The distinction he draws between abstraction in the usual sense of the word and what he calls "reflective abstraction" is applicable to the contrast between anthropological models based on thematic generalization or typology, as opposed to those based on transformation or permutation:

When a property is arrived at by abstraction in the ordinary sense of the word, "drawn out" from things which have the property, it does, of course, tell us something about these things, but the more general the property, the thinner and less useful it usually is. Now the group concept or property is obtained, not by this sort of abstraction, but by a mode of thought characteristic of modern mathematics and logic – "reflective abstraction" – which does not derive properties from *things* but from our ways of *acting on things,* the operations we perform on them; perhaps, rather, from the various fundamental ways of *coordinating* such acts or operations – "uniting," "ordering," "placing in one-one correspondence" and so on. (Piaget 1971:19)

Group models preserve the richness of data through the study of significant contrasts, instead of diluting or obliterating it through weak generalizations.

An excellent intuitive characterization of a group is given by Robin Gandy

(1973) in his essay on the concept of structure in mathematics and its empirical application:

A group can be thought of as a set of actions A, B, ..., or operations which can be compounded together – do A and then do B. The resulting action must again be a member of the group; the process of compounding is usually called "multiplication." Inaction is to be counted as a member of the group (the identity or neutral element). Each action must be invertible, so that if one compounds an action with its inverse the result is inaction – one gets back to where one started. Finally, the result of a sequence of actions may depend on the order in which they are performed, but must not depend on the order in which they are compounded. Thus "do A and then do the result of compounding B followed by C" must have the same effect as "do the compound of A followed by B and then do C" – either of these can then be simply read as "do A, then B, then C." (Gandy 1973:144–5)

"Actions" or group elements can mean arithmetic operations on numbers such as addition or multiplication (for which "inaction" would be the numbers 0 or 1), rotations or other symmetries of a body, permutations of the attributes of an object, and so on. Gandy's characterization states informally the conditions compounding of these actions must satisfy in order for there to be a group.

Groups

A *group* consists of a set $X = \{x, y, z, \dots\}$ of elements (not necessarily finite, although that is the only kind we will have occasion to use), together with an unspecified binary operation usually denoted by o as in $x \text{ o } y$ (in order to make it look unlike the conventional operations such as addition and multiplication) and called "multiplication" or "group product," satisfying the following assertions, called axioms. In all four statements, the Greek letter, ϵ means "is an element of."

A1. (Closure) If $x, y \epsilon X$, then
$x \text{ o } y \epsilon X$.
A2. (Associative law) For all $x, y, z \epsilon X$,
$(x \text{ o } y) \text{ o } z = x \text{ o } (y \text{ o } z)$.
A3. (Identity) There exists an *identity element* $i \epsilon X$ such that for all $x \epsilon X$,
$i \text{ o } x = x$ and $x \text{ o } i = x$.
A4. (Inversion) For each $x \epsilon X$ there exists an element $x^{-1} \epsilon X$ (called the *inverse* of x) such that
$x \text{ o } x^{-1} = i$ and $x^{-1} \text{ o } x = i$.

Let S be a subset of the group X. The *closure* of S is written $<S>$ and consists of S together with all other elements of X, if any, which are obtained by repeated application of the group operation $x \text{ o } y$ to elements $x, y \epsilon S$. A *subgroup* of the group X is a subset of X which is itself a group; that is, it satisfies axioms A1 through A4. It is well known in group theory that for every nonempty subset S of X, $<S>$ is a subgroup of X; it is called the *subgroup generated by* S. If $<S> = X$, then S generates X and is a *set of generators* of the group X. Continuing, S is a *minimal generating set* if $<S> = X$ but no proper subset of S *generates* X; that is, for each $s \in S$, $<S - s> \neq X$.

Example 1. Consider $X = Z_6 = \{0, 1, 2, 3, 4, 5\}$ with operation $+_6$ (addition modulo 6). The singleton $S = \{2\}$ generates $<S> = \{0, 2, 4\}$ since

$$2 +_6 2 = 4 \quad \text{and} \quad 4 +_6 2 = 0.$$

Note that this subgroup $<S>$ is isomorphic to the group Z_3.

Example 2. The Klein 4-group X can be defined by the following group table:

o	i	a	b	c
i	i	a	b	c
a	a	i	c	b
b	b	c	i	a
c	c	b	a	i

In this group, each $x \in \{a, b, c\}$ satisfies $x \circ x = i$ (every element is its own inverse), and the product of any two of these elements is the third element. Thus if we take $S = \{b, c\}$, then $<S>$ is the entire group X, so that S is a minimal generating set. In fact, any two elements other than i form such a set.

The *order of a finite group X*, written $|X|$, is the number of elements in X. And the order of an element $x \in X$ is the order of the subgroup $<x>$ generated by x. For example, the order of the Klein 4-group is 4, and the order of each of its elements $x \neq i$ is 2.

We can now define several types of groups. A *boolean group B* is finite, and every element of B other than the identity i has order 2. The *cyclic group* of order n is isomorphic to $Z_n = \{0, 1, 2, \ldots, n - 1\}$ under $+_n$; it is generated by the singleton, $\{1\}$.

A *permutation group X* is a group whose elements are permutations and whose group operation is the product of permutations, defined as the first followed by the second. Recall the description by Gandy (1973) given above, ''...do A and then do B.'' Permutations are sometimes called *transformations.* By definition, a *permutation* α takes a finite set V of *objects,* and performs a one-to-one correspondence that maps the elements (objects) of V onto themselves; α is said to *act on V*. A permutation can leave all the objects in V unchanged; this is the *identity permutation*. At the other extreme, a permutation might map each object of V onto a different one. The *degree* of a permutation group X is the number of objects in V.

The *symmetric group* S_n is the permutation group consisting of all $n!$ permutations on n objects, usually denoted by $\{1, \ldots, n\}$. Thus S_n has degree n and order $n!$. We now verify that S_2 and Z_2 are isomorphic, written $S_2 \simeq Z_2$. As $Z_2 = \{0, 1\}$ under $+_2$, its group table is

Z_2:

$+_2$	0	1
0	0	1
1	1	0

The symbol (1 2 3 4) in a permutation means that object 1 is sent onto object 2, as well as 2 onto 3, 3 onto 4, and finally 4 onto 1. This description applies to any symbol (1 2 ... n), so that in particular the permutation (1 2) sends 1 onto 2 and 2 onto 1 and thus interchanges 1 and 2; (1 2) is called a *transposition*. By definition, S_2 has $2! = 2$ permutations: the identity (1)(2) and the transposition (1 2). So its group table is

S_2:

Permutation product	(1)(2)	(1 2)
(1)(2)	(1)(2)	(1 2)
(1 2)	(1 2)	(1)(2)

The *sum*[1] of two permutation groups X and Y consists of all permutations $x\,y$ obtained from the juxtaposition of permutations $x \,\epsilon\, X$ and $y \,\epsilon\, Y$. Thus the sum $S_2 + S_2$ is isomorphic to the Klein 4-group. Similarly, every (finite) boolean group is isomorphic to a sum of n groups S_2 for some positive integer n. It is denoted by B_n and has order 2^n. Thus the three smallest boolean groups are $B_1 = S_2$, $B_2 = S_2 + S_2$, and $B_3 = S_2 + S_2 + S_2$. For the boolean group, B_3, the $2^3 = 8$ elements can be written as the binary sequences:

000 001 010 100 011 101 110 111.

The group operation of B_3 is then componentwise addition modulo 2, so that, for example, $011 + 101 = 110$. Clearly, B_3 is generated by the unit vectors 001, 010, 100, which form a minimal generating set. Of course, this generalizes at once to any B_n.

The graph Q_n, called the *n-cube*, has 2^n points that are all the binary sequences of length n, with two points adjacent whenever their sequences

[1]This operation has been called by various authors the sum of two permutation groups, their product, direct sum, and direct product.

Graphs and groups

differ in exactly one of the n places (Harary 1969:23). Thus the points of Q_n are precisely the elements of the boolean group B_n. The three smallest cubes are shown in Fig. 8.1.

Note that in Q_n, the distance between two points is the number of places in which their sequences differ. For example, 01 and 10 are points of Q_2 differing in two places and hence are not adjacent, but are at distance 2.

Klein groups

A basic underlying model for the analysis of symbolic structure in *Mythologiques* is that of a group. The Klein 4-group, for example, is evoked in the third and fourth volumes by using the symbols x, $-x$, $1/x$, $-1/x$, which stand for the operations of additive and multiplicative inverses of numbers. Since culinary symbolism plays such a central role in Lévi-

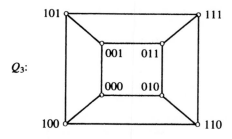

Fig. 8.1. The three smallest cubes.

Strauss's work, let us consider how it is used to articulate gender concepts but this time in a New Guinea rather than a South American society.

According to Lévi-Strauss (1966, 1968) "the cooking of a society is a language in which it unconsciously translates its structure." A basic analytical concept of this language or code is the culinary triangle, a semantic structure defined by the categories of the raw, the cooked, and the rotten. The second and third terms contrast with the first on the dimension elaborated versus unelaborated, and the second and third contrast with each other as cultural versus natural elaborations. Within this abstract triangle are concrete triangles representing particular cultural systems. For example, in some systems, corresponding to the categories raw, cooked, and rotten, are roasted (relatively unelaborated), smoked (maximally elaborated), and boiled (elaborated and metaphorically associated with the rotten, as reflected in such expressions as *pot pourri*). Actually, the systems may be more complex, giving tetrahedrons and so on. In any social group culinary modes are usually carefully distinguished and used to express or encode social and cultural differences. Thus the contrast roasted versus boiled,

157

found in innumerable cultures, may correspond to such distinctions as exo- versus endocuisine, male versus female, bush (hunting) versus village (sedentary) life, aristocrat versus commoner, life versus death.

Just as the distinction raw versus cooked is homologous with the distinction nature versus culture (humans cook their food, animals do not), so it may be used to express individual states. Culinary distinctions and operations often serve to symbolize social status and mediate transitions. Thus persons "deeply involved in a physiological process," for example the newborn, women who have given birth, and just pubescent girls, may be symbolically cooked and thereby removed from nature and socialized. Persons in an abnormal or peripheral state may be conceived of as "raw" and as a correction of excess may even be transformed into raw persons by the ingestion of certain foods. For example, in former times in certain areas of France, to ridicule celibacy "...the unmarried elder brother and sister were forced to eat a salad consisting of onions, nettles and roots or of clover and oats; this was termed 'making them eat salad,' or 'making them eat turnip'" (Lévi-Strauss 1975:335). In Oceania, there is the well-known Samoan custom of *ifonga,* the symbolic roasting of the guilty, which upon analysis turns out to be part of a system of culinarily expressed punishments for social offenses including adultery, murder, and theft (Hage 1979d). Such rites, according to Lévi-Strauss, constitute a kind of "paralanguage" that can be employed to modify a practical situation, or to characterize it, or most commonly, to perform both functions at the same time. It is apparent from *Mythologiques* that these rites are particularly associated with sexuality and marriage.

In Mountain Arapesh society as described by Margaret Mead (1938, 1940, 1947, 1961, 1963), almost all basic social categories are expressed in a culinary idiom. In kinship, the fundamental distinction between relatives linked by direct descent and same sex as opposed to those linked by opposite sex connections (Mead's "Class One" and "Class Two" categories) is correlated with payments of cooked as opposed to raw food, and the merging of alternate generations is expressed by common food taboos rather than by the kinship terminology. The moieties that regulate competitive feasting are similarly defined by the contrast raw versus cooked – in this case mythologically by the manner in which the members of each group are said to consume their meat. Spatially, the concentric dualism of the village layout is based on a complex set of oppositions which includes vegetable versus meat, raw versus cooked, and food versus refuse, as well as center versus periphery, male versus female, death versus life, and good versus bad. One of the permutation groups in this rich domain may be illustrated by means of that part of the life cycle involving marriage.

In Arapesh, the distinction between the roasted and the boiled is a basic one. According to Mead, the former is regarded as "a despised method of preparing food employed by bachelors and widowers" (Mead 1940:415), whereas the latter is esteemed and considered a privilege of marriage. In Arapesh conception, marriage is the supremely desired social state. It alone signifies legitimate sexuality, and it is preeminently defined by a wife's cooking for her husband. It is perhaps fitting that Arapesh bachelorhood is associated with the roasted, since according to Lévi-Strauss, there is an "affinity of the roast with the raw" (incomplete cooking) which is "on the side of nature."

So, on two counts, the roast can be placed on the side of nature, and the boiled on the side of culture. Literally, since boiled food necessitates the use of a receptacle, which is a cultural object; and symbolically, in the sense that culture mediates between man and the world, and boiling is also a mediation, by means of water, between the food which man ingests and that other element of the physical world: fire. (Lévi-Strauss 1968:480).

The probably universal metaphorical association between eating and sex is well known in Arapesh from a famous aphorism that equates food restrictions with incest taboos:

Other people's mothers
Other people's sisters
Other people's pigs
Other people's yams which they have piled up
You may eat,
Your own mother
Your own sister
Your own pigs
Your own yams which you have piled up
You may not eat. (Mead 1940:352)

The general aptness of the raw–cooked distinction as a symbol for the absence and presence of sexuality is suggested by linguistic connections such as those noted by Roheim (1932:77) for a Central Australian language in which "'Raw' means also not sufficiently developed for coitus, 'cooked' means nubile, 'to eat' is the usual vulgar expression for coitus."

In the Arapesh system, each term of the opposition between bachelorhood and marriage, associated with unelaborated versus elaborated food, is a term of another opposition between male sexual precocity and female reproduction. Bachelorhood enjoins sexual abstinence because premarital sex, especially for males, would endanger growth and, as is well known from *Sex and temperament* (Mead 1963), growth is the ultimate Arapesh value. An Arapesh youth is the "custodian of his own growth," which he ensures by hygienic practices, food restrictions, and chastity. In the event of sexual precocity (and we note that the etymological meaning of precocity is "pre-cooked"), there is a special ceremony for a boy:

The boy who is found guilty is punished by being made to chew a piece of areca nut that has been placed in contact with a woman's vulva, if possible with the vulva of the woman, usually his betrothed wife, with whom he has had intercourse. This ritual break of the most deeply felt taboo in Arapesh culture, the taboo that separates the mouth and the genitals, food and sex, is felt to be punishment enough; and while the guilty are punished, all are cautioned against similar indulgence. Sex is good, but dangerous to those who have not yet attained their growth. (Mead 1963:75)

If the threat to bachelorhood is premature sex, the threat to marriage is reproductive sex. Female reproductive activities are inimical to culinary ones. Menstruation is dangerous ("It is very convenient for a man to have two wives: when one is menstruating he has another to cook for him" [Mead 1963:109]), and childbirth is especially dangerous – so much so that it requires an act of self-mediation: "The wife performs a special ceremony which will ensure that her cooking will not be injured by the experience she has just passed through. She makes a [steamed] mock vegetable pudding from inedible coarse wild greens and this is thrown away so that the pigs will eat it" (Mead 1938:36).

To summarize, the culinary symbolism that expresses the "physiology" and the "pathology" of the marriage bond in Arapesh defines four states: marriage and bachelorhood (whose entire regimen prepares for marriage), which are associated with elaborated and unelaborated cuisine, respectively, and the threats to marriage and bachelorhood, namely, reproduction and premature sex, which are in turn negated by elaborated and unelaborated anticuisine, respectively. We have then the four-term system shown in Fig. 8.2, in which, through the idiom of culinary symbolism, celibacy is directly

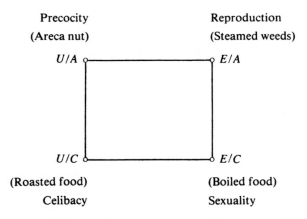

Fig. 8.2. The group B_2 represented as the graph Q_2 of culinary symbolism in the Arapesh life cycle.

[2]The "automorphism group" of the square (2-cube, Q_2) of Fig. 8.2 is known as the dihedral group D_4 having degree 4 and order 8. This is not intended here, as the first and second digits of a two-digit binary number cannot be interchanged.

160

opposed to precocity and to sexuality and diametrically opposed to reproduction, the outcome and temporary absence of sexuality.

The structure of this system is the boolean group B_2, which is the Klein 4-group as noted previously.[2] As $B_2 = S_2 + S_2$, let us say that the first S_2 acts on object set $\{a, b\}$ and the second S_2 acts on $\{c, d\}$. Then B_2 is a permutation group of degree 4 as it acts on $\{a, b, c, d\}$ and of order 4, since its elements (permutations) are: $(a) (b) (c) (d) =$ the identity, $(a) (b) (cd)$, $(ab) (c) (d)$, and $(ab) (cd)$. Now regarding each S_2 as isomorphic to Z_2 (p. 156), the four permutations in B_2 can be abbreviated as 00, 01, 10, and 11, respectively, with componentwise addition modulo 2. These permutations form precisely the set of points of Q_2 in Fig. 8.1 and hence of the Q_2 in Fig. 8.2. In Fig. 8.2 the first S_2 in $B_2 = S_2 + S_2$ consists of $\{U, E\}$ in place of $\{0, 1\}$, where $U =$ unelaborated and $E =$ elaborated, and the second $S_2 = \{C, A\}$ where $C =$ cuisine and $A =$ anticuisine. The fact that these cultural practices form a mathematical group (satisfying all four of the group axioms) shows that Arapesh thought is "logical" contra both Mead and Piaget.

Ōne of the advantages of a group model is that it accounts for practices that might otherwise be classified as anomalous. What fails to fit thematically may be intelligible structurally. Such an apparent anomaly occurs here. At Arapesh marriage, a husband prepares a special soup for his wife containing, among other things, special *wabalal* yams.

After she has eaten the soup, he takes one of the *wabalal* yams and breaks it in half. She eats half and half he places in the rafters of the house; this is the earnest that she will not treat him like a stranger and deliver him over to the sorcerers. Lest she do so, tradition provides him with part of her personality also. The piece of yam is kept until the girl becomes pregnant. *This yam meal is an incongruous piece of ceremonial, possibly borrowed from the Plainsmen.* Only the insane and the feeble-minded attempt sorcery through it. [emphasis added] (Mead 1963:95)

The half a yam is classified as exuviae or "dirt" used in sorcery. Just as cuisine is concentrically defined – an Arapesh meal consists of a vegetable center and an obligatory meat garnish – so is "anticuisine":

The word "dirt" is used in pidgin-English throughout the Mandated Territory to mean "exuviae used in sorcery practices." The Arapesh classify these exuviae into two groups; to one group, which includes parts of food, half-smoked cigarettes, butts of sugar-cane, and so on, they apply the adjective that means "external" or "outside"; to the other which includes emanations from the body that are felt to retain a close connection with the body – perspiration, saliva, scabs, semen, vaginal secretion, are included here,...they apply a different specialized term. (Mead 1963:12)

The *wabalal* rite does not appear incongruous. Rather, the practice of giving yams as external exuviae to guarantee a marriage is symmetric and inverse to that of giving semen as internal exuviae to guarantee a divorce, for according to Mead:

Only in the relationships between the sexes is sorcery used as a sanction. In the occasional cases when a man makes overtures to another's neglected wife and urges her to run away with him, he will have intercourse with her as an earnest of his honorable intentions, thus leaving her with his semen, the most potent of all dirt. If he does not fulfill his promises he is risking sorcery. (Mead 1961:43–4)

The structure of this system is, of course, another permutation group, the boolean group $B_1 = S_2$, which has degree 2 (two objects) and order 2 (two elements).

We note that the structural intelligibility of anti-yams in particular and of anticuisine in general as a sign of actual or potential social disjunction in Arapesh thought is further exemplified in myths about the passage from nature to culture (Myth 12, Mead 1940; Myth XXV, Fortune 1942): Originally men lived alone and ate rotten woodchips = "our yams" until Sharok, the Cassowary Woman (just like Star Woman in the South American myths [Lévi-Strauss 1975]), descended to earth and introduced them to real food (cooked yams), domesticity, and the arts of culture. In the myths, the anti-yams = rotten woodchips = vegetable food are opposed to hardwood = meat (Myth XXX, Fortune 1942). This offers a striking parallel to certain South American myths in which culture is similarly prefigured by a culinary antiworld.

"Multiplicative groups"

Since we are dealing with algebraic groups, it is natural to consider their implications for Piaget's theories about the structure of primitive thought, especially since these theories have now entered the mainstream of anthropological discourse (Hallpike 1976, 1979). Piaget distinguishes operational from preoperational or intuitive thought. Operational thought satisfies the axioms of a group, whereas preoperational thought falls short in two respects. It lacks an inverse, which means that one cannot conserve an object when it has been transformed, and it is not associative, which means that transformations, when they finally do occur, cannot be combined; nor can one recognize that the same result can be obtained by alternative routes. Intuitive thought is dominated by perception and is immobile, consisting of rigid unanalyzed configurations. It is "centered," which means that an object is seen only in relation to oneself and not in relation to other objects:

The mechanism of this type of reaction is easy to unravel: the subject finds no difficulty in concentrating his attention on the whole B, or on the parts A and A', if they have been isolated in thought, but the difficulty is that by centring on A he destroys the whole, B, so that part A can no longer be compared with the other part A'. So there is again a non-conservation of the whole for lack of mobility in the successive centralizations of thought. (Piaget 1960:133)

In his discussion of anthropology, Piaget maintains that primitive thought is preoperational or just possibly limited to certain concrete operations. Although he concedes that kinship systems "bear witness to a much more advanced logic," these he regards as "already crystallized cultural products to be distinguished from the everyday reasoning of the individual." However, in making this distinction, he questions whether even the former are not the "outcome of function producing applications rather than groupings in the operational sense" (Piaget 1971:115–16); that is, trial-and-error approximations that are combinations only resembling group structures.

Piaget's theories provide the basis for C. R. Hallpike's recent book *The foundations of primitive thought,* where they are combined with a kind of empiricism according to which binary oppositions in thought are due to the "twoness of reality" rather than to mind (as though mind were not a part of reality). In Hallpike's view, collective representations may consist of congeries of binary oppositions, but these are in general not organized into coherent systems; they only appear to be because of the literate anthropologist's predilection for arranging things in tables. Hallpike claims to have "shown in detail that primitive thought generally conforms to Piaget's criteria of the advanced stages of pre-operatory thought at the level of collective representations" (Hallpike 1979:489). But a system like the one in Fig. 8.2 is based on binary oppositions that do combine to form a structure satisfying the axioms of a group. In fact, this system is exactly equivalent to what Piaget (1957) calls a "multiplicative group"; that is, one based on the product or combination of classes, as when differences in material, brass-steel (A_1/A_1'), combined with differences in length, short-long (A_2/A_2'), give a complex classification of objects.[3] Such a grouping combines classes in all possible ways, as in

$$(A_1 + A_1')(A_2 + A_2') = A_1A_2 + A_1A_2' + A_1'A_2 + A_1'A_2' \, ,$$

and is a prime instance of concrete operational thought. However, this may be the result of our inclination to put things in cubes, so we will use another example, one analogous to the problem-solving tasks in Piagetian experiments. Since these experiments have been severely criticized for overlooking the effect of content on the type of mental operations expressed in the performance of a task (see the review by Shweder 1982), this example also suggests the kind of naturalistic observations that would actually be meaningful in cross-cultural studies of cognition.

One of the aspects of Micronesian navigational knowledge described in

[3]In mathematics a "multiplicative group" is a group (X,o) whose group product is multiplication. In an "additive group," it is addition.

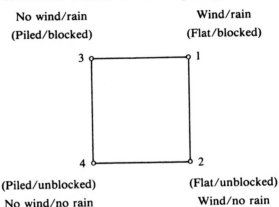

No wind/rain   Wind/rain

(Piled/blocked)   (Flat/blocked)

(Piled/unblocked)   (Flat/unblocked)

No wind/no rain   Wind/no rain

Fig. 8.3. The structure of a Micronesian sea sign.

Chapter 2 is the use of sea signs in weather forecasting. The following system was described to David Lewis by a Gilbertese navigator.

"The time to set out on a voyage depends upon the sea signs,... In considering sea signs we will begin with crabs," he continued. "We will study only the kind of crab that digs his hole straight down." He paused to demonstrate just above the tideline. "This kind of crab may do one of four things.

"He may block the mouth of the hole and scratch the sand down flat across the opening leaving marks like the sun's rays. This means there will be wind and rain within two days.

"When the crab levels his pile of excavated sand but does not cover the mouth of the hole, there will be strong wind but no rain.

"If he just blocks his hole but does not scrape the mound flat with scratch marks, there will be rain but no wind.

"When the crab leaves the excavated sand piled up in a mound and the opening of his hole unblocked, the weather will be fine." (Lewis 1978:124–5)

We can use the system shown in Fig. 8.3 to illustrate the "complete combinatorial character" of operational thought. When wind or no wind is represented by 1 or 0, and similarly with rain or no rain, the labels of Fig. 8.3 become identical with those of Q_2 in Fig. 8.1. The multiplicative group illustrated above with $(A_1 + A_1') \times (A_2 + A_2')$ can also be thought of as a system of propositions. According to Piaget:

If we make the proposition p correspond to A_1, the proposition q to A_2, the proposition $\bar{p}$ to A_1' and the proposition $\bar{q}$ to A_2', the multiplication $B_1 \times B_2$ then corresponds to:

Classes: $(A_1 + A_1') \times (A_2 + A_2') = A_1A_2 + A_1A_2' + A_1'A_2 + A_1'A_2'$

Propositions: $(p \lor \bar{p}) \cdot (q \lor \bar{q}) = (p \cdot q) \lor (p \cdot \bar{q}) \lor (\bar{p} \cdot q) \lor (\bar{p} \cdot \bar{q})$

Product number   1   2   3   4

164

Propositional operations are thus constructed simply by combining n by n these four basic conjunctions. The 16 binary operations of two-valued propositional logic therefore result from the combinations given below (written in numerical form):

0; 1; 2; 3; 4; 12; 13; 14; 23; 24; 34; 123; 124; 134; 234 and 1234.

Elementary groupements are distinguished from the higher groupements which form the system of propositional operations by the fact that the latter is based upon a combinatorial system. Elementary groupements have not yet a complete combinatorial character. For example, multiplicative groupements of classes or relations are solely based on the multiplication of elements 2 by 2 or 3 by 3, etc., but not on combinations among the resultant products (1 to 4 or 1 to 9, etc.), as in the case of the 16 binary propositional operations formed from the products 1 to 4. Another way of expressing the fundamental difference between the two kinds of structures, is to say that elementary groupements are based only on simple sets (the included classes A>B>C, etc.) or on product sets (the multiplicative classes A_1A_2; A_1A_2'; etc.), whilst propositional structures are based on what is called in the theory of sets a *set of all sub-sets,* combinations taken n by n among the product sets. (Piaget 1957:30–1)

The combinatorial operations of propositional thought are shown in problem-solving experiments in which the subject varies one factor at a time to arrive at a solution: For example, rods of different metals and cross-sections are classified and then analyzed to determine what produces bending (Inhelder and Piaget 1958). It seems unlikely in the extreme that a Gilbertese navigator would not mind his *p*s and *q*s sufficiently to deduce that subset {1, 3} means rain, subset {1, 2} wind, and so on,[4] whether this happens to be an item of practical truth or an instance of the intellectual razzle-dazzle that characterizes these systems of knowledge. It is easy to imagine such an individual failing a psychological test designed to determine his skill at combinatorics, because the stimulus materials are utterly novel, because the instructions are poorly communicated, or because of plain lack of interest in performing a culturally meaningless or silly task. As R. H. Barnes concluded in his study of Kédang number use (a counterexample to Piagetian theory): "We might expect an anthropology of cognitive development to investigate just what capacities are required to build a house, navigate a ship, operate an alliance system or compete effectively in the marketplace" (1982:20).

So far we have considered only the boolean group B_2 as modeled by the graph Q_2. This particular group does, however, play a central role in structuralism, for it is not only the Klein 4-group of *Mythologiques,* but it is also the *Viergruppe* or INRC group of genetic epistemology (Piaget 1957). And it is the "semantic square" or "constitutional model" of semiology

---

[4]In his pioneering study of switching theory, Claude Shannon (1938) modeled the set of all on-off switching patterns with boolean functions, each regarded as some subset of the points of the cube Q_n.

(Greimas 1970). But by its very nature Q_2 always has the potential of being but one facet of a larger structure. That is, any boolean group B_n is a component of the next larger group B_{n+1}, so that any symbolic system may turn out to be a subsystem of a larger one. Thus analysis may stop short by considering only single binary dimensions or their union, as in lists given in tables, or by considering only Klein groups or semantic squares. Thus each face of Q_3 is Q_2, so that some particular Klein group may be one of six subgroups conjoined in a larger and coherent structure. For example, all the foods in the Arapesh system, as we have defined it, are vegetable. The Arapesh, however, make a fundamental distinction between vegetables and meat, the supremely valued food.[5] As has been shown elsewhere (Hage and Harary 1983), the system in Fig. 8.2 is in fact part of a larger one defined by three unit vectors that generate B_3. This hierarchical and nested nature of boolean groups allows for the recognition and treatment of increasingly complex structures.

Larger groups

In *Primitive classification*, Durkheim and Mauss comment on the natural relation between divination and classification:

A divinatory rite is generally not isolated; it is part of an organized whole. The science of the diviners, therefore, does not form isolated groups of things, but binds these groups to each other. At the basis of a system of divination there is thus, at least implicitly, a system of classification. (Durkheim and Mauss 1963:77)

Divinatory systems are among the most obviously combinatorial of all systems of classification, perhaps because of the physical apparatus – bones, coins, sticks, the behavior of animals – with which they are associated. Some, such as Micronesian sea signs, are small in structure and narrowly restricted in application, whereas others are large, sometimes very large, and open to broad interpretation. The Kalanga "divinatory calculus" described by Werbner (1973), for example, has as its underlying structure the group B_4 and is used in domestic séances for the interpretation of all manner of misfortune. One of the most comprehensive and best known systems is the *I Ching or Book of changes,* which is based on an all-embracing set of symbolic correspondences. It provides a convenient example of a group larger than B_2, the Klein 4-group. It also illustrates a relation between mathematical and symbolic complementarity and an isomorphism between primitive classification and modern mathematics.

The *I Ching* consists of 64 hexagrams (*kua*) resulting from all possible permutations of two kinds of lines taken six at a time. The lines are *yin* (female) = broken and *yang* (male) = solid. They probably originally repre-

[5]The consumption of meat in any significant quantity is the occasion for a feast, as opposed to the primarily vegetable domestic meal.

sented short and long rods used in divination. Each hexagram is named, defined by various social and natural properties, and provided with a commentary for purposes of prognostication and explanation.

According to Needham (1956), the *I Ching* most likely began as a collection of peasant omen texts that became associated with stick divination, and gradually through accretion came to classify the universe. To forecast the future, one selected randomly one of the hexagrams and interpreted its associated commentary. To explain natural phenomena, one referred them to their proper *kua,* which described their essential character. Although Durkheim and Mauss were unable to account for its social origin (in a clan system), as in their interpretations of Australian and American Indian systems of correspondence, Needham views the *I Ching* as a "mirror image of Chinese bureaucratic society, as a kind of administrative approach to nature" in which an event is dealt with not in causal terms, but by referring it to the appropriate department.

The hexagrams are composed of two trigrams (also called *kua*). The trigrams, like the hexagrams, are named and form the basis of a large system of cultural and natural correspondences which, as shown in Table 8.1, are similar to certain of those described in *Primitive classification*. The trigrams themselves can be arranged in various ways. Fig. 8.4(a) (from the Wilhelm and Baynes translation, 1967) shows the *Fu Hsi* (Sequence of Earlier Heaven or Primal Arrangement). Here the eight trigrams are ordered by compass direction and by opposing pairs that are formally and symbolically complementary. Fig. 8.4(b) shows the *King Wen* (Sequence of Later Heaven, or Inner-World Arrangement). Here the trigrams are ordered by cyclic seasonal progression.

Another way to view the trigrams, one that brings out very clearly their logical genesis and structure, is as a 3-cube. This was discovered by Z. D. Sung (1934) and described by Martin Gardner (Fig. 8.5):

Z. D. Sung tells how he was rotating a matchbox in his hand one day (to simulate the earth's rotation as it goes around the sun) when he suddenly perceived a natural way to generate the eight trigrams at the corners of a cube.

Let the three Cartesian coordinates of a unit cube, x, y, z indicate the first, second and third digits of a three-digit binary number. Label the corner where the coordinates originate with 000. The other corners are labeled with three-digit binary numbers for 0 through 7, with 0 and 1 indicating the distance of the corner from the origin in each coordinate direction. The eight numbers correspond, of course, to the eight trigrams, with complementary trigrams at diametrically opposite corners of the cube. By a similar procedure corners of unit hypercubes generate the higher-order polygrams. The 64 hexagrams correspond to six-digit binary numbers at the corners of a six-dimensional hypercube. (Gardner 1974:110)[6]

[6]A *hypercube* is a cube graph Q_n with $n \geq 4$. Similarly, in mathematical usage, a hypersphere requires a euclidean dimension greater than 3, and so forth. Here a trigram means a binary number with three digits; the eight trigrams in each of the three diagrams of Fig. 8.4 are precisely the same as the points 000 to 111 of the cube (not hypercube) Q_3 in Fig. 8.1.

Table 8.1. *Symbolic correspondences of trigrams in I Ching*

Trigram	Name	Direction	Family relation	Season	Body part	Natural object	Animal	Color
☰	*Chhien*	S (NW)	Father	Late autumn	Head	Heaven	Dragon, horse	Deep red
☷	*Khun*	N (SW)	Mother	Late summer, early autumn	Abdomen	Earth	Mare, ox	Black
☳	*Chen*	NE (E)	Eldest son	Spring	Foot	Thunder	Galloping horse, flying dragon	Dark yellow
☵	*Khan*	W (N)	Second son	Midwinter	Ear	Moon and freshwater lakes	Pig	Blood red
☶	*Ken*	NW (NE)	Youngest son	Early spring	Hand and finger	Mountain	Dog, rat	—
☴	*Sun*	SW (SE)	Eldest daughter	Late spring, early summer	Thigh	Wind	Hen	White
☲	*Li*	E (S)	Second daughter	Summer	Eye	Lightning (and sun)	Pheasant, toad, crabs, snail, tortoise	—
☱	*Tui*	SE (W)	Youngest daughter (concubine)	Midautumn	Mouth and tongue	Sea and seawater	Sheep	—

Source: Needham (1956).

Graphs and groups

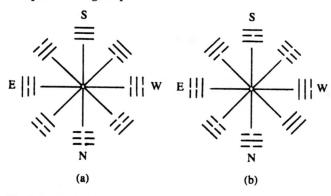

Fig. 8.4. Arrangements of trigrams in the *I Ching*.

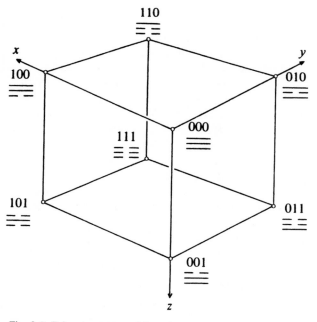

Fig. 8.5. Trigrams generated by a 3-cube (from Gardner 1980).

Interpreting the *yang* (solid) lines as 0 and the *yin* (broken) lines as 1, the structure of the trigrams and hexagrams is isomorphic to that of binary arithmetic. Leibniz, who discovered binary arithmetic in the seventeenth century, and who was something of a sinologist, was astounded when he came across this system – in particular when he saw the hexagrams arranged in a sequence corresponding to 0 to 63. Although it cannot be said that the Chinese consciously developed and applied binary arithmetic, as Leibniz

169

believed, they certainly anticipated it, and according to Needham (1978), it encouraged Leibniz to work on the first computing machines. This is a clear-cut case in which *la pensée sauvage* embodies the same principles as those that underlie such modern sophisticated systems as may be found in everything from digital computers to the all-or-none models of the nervous system (see the excellent discussion in Needham 1956, 1978).

Group models enable us to study symmetry in empirical structures. They may improve ethnographic interpretation, and they are directly pertinent to evaluating propositions concerning the "operational" status of primitive thought. We have explicated boolean groups, which are fundamental in classification systems. There are, of course, numerous other types of groups that may have application to such diverse topics as the "reflection" and "rotation" in pottery styles or the physical movements in ritual.[7]

Conclusion

Much of anthropology is concerned with elucidating the logic that underlies sets of social and cultural relations. This requires a metalanguage that is descriptively adequate and analytically rich. Our objective has been to show that a branch of topology and combinatorics known as graph theory provides a family of models for an exact and imaginative treatment of the most diverse kinds of empirical structures. Graphs, and the interactions between graphs and matrices, relations, duality laws, and groups, enable the study of such phenomena as reachability in social networks, efficiency in cognitive schemata, centrality in spatial layouts and exchange systems, consistency in kinship norms and alliance systems, logic in social relations, rank in political hierarchies, productivity in subsistence activities, transformations in myths, and permutations in symbolic structures. Graph theory makes it possible to take seriously and explicate rigorously such fundamental notions as "social relation" and "cultural logic." Graph theory is implicit in all fields of anthropology. Used explicitly, it provides clear, natural, and powerful models for understanding scientific as well as native concepts and for expanding the range of structural analysis.

[7]A geometric approach to the analysis of symmetry in ceramic and art styles is given in Washburn (1977) and Crowe (1975).

Appendix: Axiomatics

Thus mathematics may be defined as the subject in which we never know what we are talking about, nor whether what we are saying is true.

Bertrand Russell,
"Recent work on the principles of mathematics"

The order of presentation in this book has been historical rather than rigorously logical. It is only now, after we have introduced the particular types of models, that we define their foundations and discuss the general concept of a mathematical model. In the course of this discussion we will develop the axiomatic basis of the concepts relation, network, and group, and define the equivalence of two models, called isomorphism, and the simplification of one model by another, called homomorphism.

Every utilization of mathematics in the preceding chapters as a theoretical framework for topics in anthropology has been an instance of a mathematical model, which is now defined precisely. For this purpose, we must develop the notions of an axiom system and of a model of an axiom system.

Axiom systems

The quote from Bertrand Russell that begins this chapter is explained by the following definition of an axiom system. In order to avoid circular definitions, in which ultimately a concept is defined in terms of itself, certain basic terms called *primitives* must be deliberately left undefined, and then all other concepts are constructed on this definitional base. Similarly, in order to avoid circular reasoning in which a statement is utilized as a reason for its own validity, certain basic assertions called synonymously *axioms* or *postulates* are assumed to be true without proofs.

Formally, an *axiom system* Σ is an ordered pair of sets (P, A) where P is a set of undefined terms or *primitives* and A is a set of unproved assertions or *axioms,* taken to be true.

Surely the best-known axiom system is that of Euclid for plane geometry in which $P = \{$ point, line $\}$ and A consists of his axioms, the most celebrated being Euclid's Parallel Postulate:

Appendix

> Given a line L and a point A not on L, there exists another line L' containing A which has no points in common with (is parallel to) L.

By a *mathematical system* is meant an axiom system Σ together with the set T of all the theorems that can be derived from the axioms in A about P, and about so-called higher-order concepts defined in terms of the primitives. A *theorem* may be regarded as an implication of the form h (the hypothesis) implies c (the conclusion), customarily written $h \Rightarrow c$, which has been proved. To prove (or derive or demonstrate) a theorem, one must more or less formally make a list of statements and reasons, in which the final statement is c, the conclusion, and the various types of admissible reasons include the hypothesis h, an axiom, a previous theorem, a definition, a previous step in the proof, or a law of logic.

Models of an axiom system

Our description of this topic is based on that in Wilder (1952). A Σ-*sentence* is a statement concerning the primitives of Σ that is either true or false. A *model M* of an axiom system $\Sigma = (P, A)$ is an assignment of meanings to P that makes each of the Σ-sentences in M become a true statement. A model is also called a *realization* of Σ.

An illustration is provided by taking an axiom system for a graph, as follows:

$\Sigma = (P, A)$ defines a *graph* when $P = \{V, E\}$ where V is a set of points, E is a set of lines, and A is the following set of axioms.

A1. Set V is finite and not empty.

A2. Set E is a subset of the set of all two-element subsets of V. Thus each element e in E is a set of two distinct elements of V. (In other words, every line consists of two different points.)

It follows from axiom A2 of this simple axiom system[1] that no line joins a point to itself, because each line consists of two different points, and that no two distinct lines join the same pair of points, because E is a set and no set contains any element more than once. When $e = \{u, v\}$, we say that line e joins points u and v, which are *adjacent points*.

Two models M_1 and M_2 of Σ are called *isomorphic* if there is a one-to-one correspondence between the sets in M_1 and M_2 that "preserves all Σ-sentences." This will become clearer after the illustration for graphs below. We give an example by showing two different-looking but isomorphic graphs in Fig. A.1, in each of which every point is adjacent just to all

[1] Veblen (1922) defined an n-dimensional simplicial complex in this manner and discussed graphs as 1-dimensional simplicial complexes.

172

Axiomatics

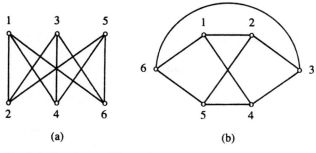

Fig. A.1. Two isomorphic graphs.

points of opposite parity (odd or even). Both graphs are presentations of the complete bipartite graph $K_{3,3}$.

An *automorphism* of a model M of an axiom system Σ is an isomorphism of M with itself. It is well known that the set of all automorphisms of M necessarily forms a group, variously called the *automorphism group* of M or the group of symmetries of M or most briefly, frequently, and simply, *the group* of M.

To illustrate automorphisms and the automorphism group of a model M of an axiom system Σ, we take $\Sigma = (\bar{r}, s)$, so that we are dealing with graphs, and we consider M as the random graph G of Fig. 2.4. The set V of four points of G is then the set Q of objects in Chapter 8, and each automorphism of G is a permutation acting on V. Clearly the two points of degree 2 in G can be left unchanged or can be interchanged. And the same holds for the two points of degree 3 in G. Hence the (automorphism) group of G is precisely the sum $S_2 + S_2$.

Much has been written in the social science literature concerning the next type of mapping from one model M of an axiom system Σ onto another model M' of Σ. We have already shown a number of ways in which the graphical structure of an empirical relation can be simplified, including the condensation of a digraph with respect to its strong components, the clustering of a signed graph, and the cliques, connected components, and blocks of a graph. Now we can define precisely a general concept for simplifying a model M of Σ that has important applications in all of mathematics, and thus in particular for graphs.

A *homomorphism* of M_1 onto M_2 is a mapping of the sets of M_1 onto those of M_2 that preserves all Σ-sentences. Because such a mapping is permitted to send more than one element of M_1 onto the same element of the image model M_2, the latter is usually smaller than M_1. By definition, an isomorphism is the special case of a homomorphism in which M_1 and M_2 are equinumerous.

For graphs, there is just one Σ-sentence (!), namely,

173

Appendix

Two points u and v are adjacent; that is, joined by a line.

Thus an isomorphism between two graphs $G_1 = (V_1, E_1)$ and $G_2 = (V_2, E_2)$, where $M_1 = G_1$ and $M_2 = G_2$, is a 1 to 1 correspondence between V_1 and V_2 that preserves adjacency (Σ-sentences); similarly, a homomorphism of G_1 onto G_2 is a mapping of V_1 onto V_2 that preserves adjacency. We have already seen an example of graph homomorphism, which we now specify as such. Fig. 6.3 shows a graph G with three sets of points $S = \{1, 2, 3\}$, $T = \{4, 5\}$ and $U = \{6, 7\}$ in which two points are adjacent if and only if they are in different sets. This complete multipartite graph $G = K(3, 2, 2)$ is homomorphic to the complete graph (triangle) K_3 with point set $\{S, T, U\}$. To check that this map $K(3, 2, 2) \rightarrow K_3$ is indeed a homomorphism, two types of statements need verification for all possible pairs of points of G as now illustrated.

1. Points 1 and 4 are adjacent in G implies that their image points S and T are adjacent in K_3.
2. Points 1 and 3 are not adjacent in G implies that their image points S and again S are not adjacent in K_3 (as no point of a graph can be adjacent to itself).

We now give an equivalent formulation of graph homomorphism in terms of a coloring of its points. A *coloring* of a graph G is an assignment of one of c colors denoted by $1, 2, \ldots, c$ to each point so that no two adjacent points have the same color. A path P_8 with its eight points alternately colored 1 and 2 is shown in Fig. A.2(a). This point coloring determines a homomorphism of P_8 onto K_2. On the other hand, Fig. A.2(b) assigns four colors to $V(P_8)$ in such a way that for each pair of distinct colors i and j in $\{1, 2, 3, 4\}$, there is a line of P_8 whose two points have these colors. This

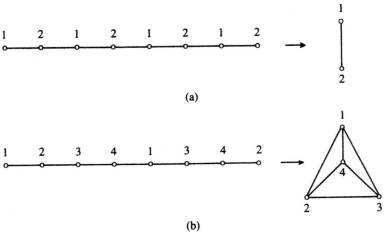

(a)

(b)

Fig. A.2. Two complete homomorphisms of a path.

174

coloring thus precisely determines a homomorphism of P_8 onto K_4 with the mapping given by the coloring of $V(P_8)$.

The *chromatic number* $\chi(G)$ of a graph G is the minimum number n such that n colors suffice for a coloring of G. Equivalently, $\chi(G)$ is the smallest n such that there is a homomorphism of G onto K_n. For many years the Four-Color Conjecture stated that every planar graph G has chromatic number at most 4. Then, with substantial and crucial computer assistance, Appel and Haken (1977) settled it in the affirmative (see the definitive book on the subject by Kainen and Saaty, 1976, for a full account of the conquest of this problem). To this day, the postage meter of the Department of Mathematics of the University of Illinois proudly proclaims: FOUR COLORS SUFFICE. It is now known as the Four-Color Theorem. (The saddest title of a mathematical research paper is that of Allaire, 1980, "Another proof of the Four Color Theorem.")

Mathematical models

"A mathematical model is an abstraction of an empirical phenomenon" is precisely the same statement as "An axiom system is realized by an empirical model." The mnemonic that applies is "A model is always the other fellow's system." To a logician, a model always is an empirical realization of an axiom system Σ. Strictly speaking, an *interpretation* of Σ is an assignment of meanings to the primitives in P that makes every axiom in A become a true statement. Then a model M of Σ is the result of an interpretation. Conversely, the axiom system Σ is then called (not by mathematicians but by social scientists) a *mathematical model* of M.

We illustrate without loss of generality with a graph. Consider a model M consisting of four people and a relation that tells when a pair of persons in this small social group is on speaking terms. In this group (not a mathematical group even if they should be a set of mathematicians), let us stipulate that exactly one pair of persons is not on speaking terms. Then the graph, which is a mathematical model of this empirical situation, is the so-called random graph K_{4-e} obtained from the complete graph K_4 by removing any edge e; this graph is shown in Fig. 2.4.

Next, recall the axiom system for groups stated in Chapter 8. We present the smallest models of this axiom system by showing the group tables of all the groups of *order n* = 1, 2, 3 and 4 elements.

The group with one element (of order 1) must consist only of the identity element i by axiom A3, which guarantees its existence, so the entire group table simply says that $i \circ i = i$. Thus its group table is

o	i
i	i

Appendix

This obviously satisfies all four axioms for a group and is quite trivial. In fact, the one-element group is technically called the *trivial group*. It is for this reason that k_1, the one-point graph, is called the *trivial graph*.

One group with two elements has $X = \{+1, -1\}$, and the group product is ordinary multiplication. Thus the group table, which tells the element $x \circ y$ for each pair x and y, is

$\circ$	$+1$	-1
$+1$	$+1$	-1
-1	-1	$+1$

From this we see at once that $+1$ is the identity element and -1 is its own inverse, since as usual $(-1) \circ (-1) = +1$.

Another group table for a two-element group has $X = \{0, 1\}$ and the group product is $+_2$, addition modulo 2:

S_2:

$+_2$	0	1
0	0	1
1	1	0

This group table has already been shown in Chapter 8. Here 0 is the identity and 1 is its own inverse, as $1 +_2 1 = 0$. These two groups are isomorphic under the one-to-one correspondence $+1 \leftrightarrow 0$, $-1 \leftrightarrow 1$. For this reason it is customary to say that up to isomorphism, there is just one group of order 2.

The same statement is true for the group of order 3 whose group table may be written in either of the following two equivalent ways, in which $w = (-1 + i\sqrt{3})/2$ and $w^2 = (-1 - i\sqrt{3})/2$ are complex numbers ($i^2 = -1$) with $w^3 = 1$:

$+_3$	0	1	2
0	0	1	2
1	1	2	0
2	2	0	1

$\circ$	1	w	w^2
1	1	w	w^2
w	w	w^2	1
w^2	w^2	1	w

These are obviously isomorphic: The first is the cyclic group Z_3 of Chapter 8, so the second is another presentation of Z_3.

But when we get to order 4, we find there are now two different (nonisomorphic) groups. The first of these is the *cyclic group* with group table either of:

Z_4:

$+_4$	0	1	2	3
0	0	1	2	3
1	1	2	3	0
2	2	3	0	1
3	3	0	1	2

or

$\times$	1	i	-1	$-i$
1	1	i	-1	$-i$
i	i	-1	$-i$	1
-1	-1	$-i$	1	i
$-i$	$-i$	1	i	-1

Axiomatics

In the latter table i stands for the familiar imaginary number which satisfies $i^2 = -1$ and $\times$ indicates the usual multiplication of complex numbers. The second is the so-called Klein-4 group $S_2 + S_2$ written either abstractly with $X = \{i, a, b, c\}$ or vectorially with $X = \{00, 10, 01, 11\}$ and $+_2$ acting componentwise:

o	i	a	b	c
i	i	a	b	c
a	a	i	c	b
b	b	c	i	a
c	c	b	a	i

$+_2$	00	10	01	11
00	00	10	01	11
10	10	00	11	01
01	01	11	00	10
11	11	01	10	00

Relations

A *sequence of length n* is an ordered list $(a_1, a_2, \ldots, a_n)$ of n elements taken from a given set V. When V is the set of real numbers and $n = 3$, each sequence (a_1, a_2, a_3), called an *ordered triple,* determines a point in the usual XYZ 3-dimensional euclidean space. In the smallest instance when $n > 1$, we have *ordered pairs* (a_1, a_2). A *relation R* is defined as a collection or set of elements each of which is an ordered pair. This concept and its terminology was proposed by the philosopher Charles Peirce, who was considering V as a set of people and hence the relation R as an interpersonal relation, such as one of the kinship relations: mother, son, uncle, cousin. In this formulation, the relation "mother" is the set of all ordered pairs (a_1, a_2) of persons such that a_1 is the mother of a_2, and similarly for other interpersonal relations. In the case that V is the set of real numbers, the relation "greater than," written as usual $>$, is the set of all ordered pairs (a_1, a_2) such that $a_1 - a_2$ is positive, that is, $a_1 > a_2$. With this background, we can now present various properties of relations in order to use them as axioms for several kinds of structural models.

In Chapter 4, we have already axiomatized various kinds of relations. We now enlarge on that discussion.

The *cartesian product* $V \times V$ of a set V with itself is the collection of all ordered pairs (u, v) where both u and v are in V. We say that R is a *relation on the set V* if R is a subset of $V \times V$. The *universal relation* on V is the entire set $V \times V$. It is called universal because it is universally true that (u, v) is in it for every choice of u, v in V, including $v = u$. When (u, v) is in relation R, it is often convenient to write uRv.

The seven basic properties of relations introduced in Chapter 4, $r, \bar{r}, s, \bar{s}, t, \bar{t}, c$, can now be regarded as individual axioms. Combinations of these seven symbols will serve as axiom systems for 12 different types of structures. *In all of these it is understood that the primitives are a set V of elements and a subset R of $V \times V$, and that the axioms include the assertions:*

Appendix

1. V is finite,
2. V is not empty (but R is permitted to be the empty set).

When there are no other axioms, this constitutes a simple axiomatization of a relation. The seven properties of relations are:

Symbol	Adjective	Definition
r	reflexive	For all u in V, uRu.
$\bar{r}$	irreflexive	For each u in V, (u, u) is not in R.
s	symmetric	For any two different elements in u, v in V, if uRv then vRu.
$\bar{s}$	asymmetric	For any two different elements u, v in V, if uRv then it is not true that vRu.
t	transitive	Whenever u, v, w are distinct elements of V, if uRv and vRw then uRw.
$\bar{t}$	intransitive	Whenever u, v, w are distinct elements of V, if uRv and vRw, then uRw is false.
c	complete	For $u \neq v$ in V, we have uRv or vRu or both.

We can now list axiom systems for 12 kinds of structures, as promised.

Structure	Axiom
1. Digraph	$\bar{r}$
2. Graph	$\bar{r}, s$
3. Complete graph	$\bar{r}, s, c$
4. Parity relation	$\bar{r}, s, t$
5. Oriented graph	$\bar{r}, \bar{s}$
6. Tournament	$\bar{r}, \bar{s}, c$
7. Partial order	$\bar{r}, \bar{s}, t$
8. Complete order	$\bar{r}, \bar{s}, t, c$
9. Equivalence relation	r, s, t
10. Universal relation	r, s, c
11. Similarity relation	r, s
12. Antiequivalence relation	$\bar{r}, \bar{s}, \bar{t}$

These structures and their construction are vividly illustrated in Fig. A.3, in which a descending line from one structure (say, digraph) to another (say, graph) indicates that exactly one more axiom (here, s) has been added to the first to define the second. Every graph is thus a digraph but not conversely, and so forth for other pairs of structure types.

The 13 diagrams of Fig. A.4 illustrate first a relation R with none of these six properties, and then each of the 12 structures listed above. We have already seen most of these structures. Chapters 4 and 2 are devoted to digraphs and graphs. A complete graph with more than two points first occurs in Chapter 2. Parity and equivalence relations are discussed in Chapter 4, as are oriented graphs and tournaments, and partial and complete orders. All these types of mathematical models are generally over-simplifications of real anthropological phenomena because their arcs (ordered pairs) are either absent or present, but do not show intensity or

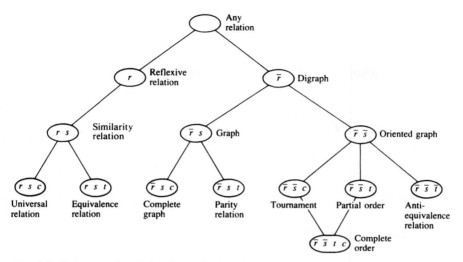

Fig. A.3. Fragment of a relational organization chart.

strength of the relationship (arc). This is done by networks, which we now axiomatize.

Networks

Familiarity with the set of real numbers must be assumed. A network can be defined on any type of structure R in the preceding section by adding the following additional postulate, which defines real networks.

Network axiom. For each arc $x = (u, v)$ in R, there is an associated real number $f(x)$.

The letter f comes from the word "function." The number $f(x)$ is called the *value* of arc x in the network, which is denoted by $N = (R, f)$. The values which $f(x)$ may attain serve to specify the type of network at hand. Two examples already taken up in preceding chapters are these:

a. Signed graphs (and digraphs) in which for each line (or arc), the value is either $+1$ or -1.
b. Markov chains. Here the underlying structure is an arbitrary relation R and the following two stipulations hold.

Markov axiom 1. For each arc x, the value of $f(x)$ satisfies the inequalities $0 < f(x) \leq 1$.

(The number $f(x)$ stands for the probability that arc X is present. Thus this axiom says that each arc in R is *possible*. There is a rapidly growing area of

179

Appendix

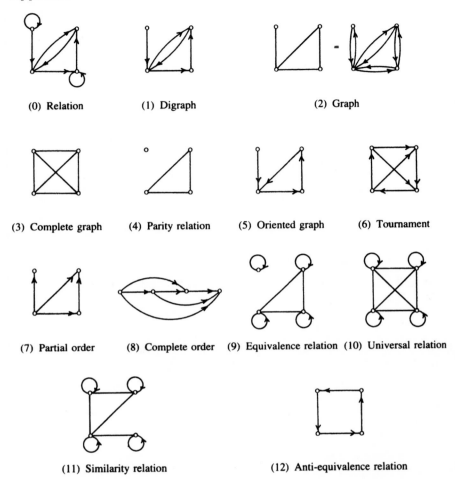

(0) Relation (1) Digraph (2) Graph

(3) Complete graph (4) Parity relation (5) Oriented graph (6) Tournament

(7) Partial order (8) Complete order (9) Equivalence relation (10) Universal relation

(11) Similarity relation (12) Anti-equivalence relation

Fig. A.4. Twelve types of structural models.

both theory and application, called "fuzzy set theory" that includes "fuzzy graph theory," which is based on this probabilistic condition.)

Markov axiom 2. For each point v of N, with outdegree d, let the arcs from v be $x_1, x_2, \ldots, x_d$ and let $p_i = f(x_i)$. Then $p_1 + p_2 + \ldots + p_d = 1$. Note that one of these arcs might be the loop (v, v).

A third kind of network involves two special points denoted usually by s and t and called the source and the sink (or terminal).[2] These occur in

[2]With the wealth of concepts necessary for the study of graphs and networks, it is really impossible to avoid the two sins of notation mentioned earlier. In Chapter 7, the source was a transmitter denoted by t and the terminus or sink was a receiver r.

180

electrical engineering and operational research, and one appears in Chapter 7.

A *2-terminal network* consists of a connected graph G, a real valued function f on $E(G)$, and two distinct points s and t distinguished from the other points. Frequently in this context the real numbers $f(e)$ are restricted to positive integers.

Groups

We have already utilized groups above. Now we present first the most standard axiom system for a group, and then some variations.

A *group* satisfies the following axiom system:

Primitives
1. A set of S of elements $a, b, c, \ldots$.
2. A binary operation o which can act on two elements, a, b of S, to give their product, $a \circ b$.

Axioms
A1. (Closure) If a, b are in S, then their product $a \circ b$ is an element of S.
A2. (Associativity) If a, b, c are in S, then $a \circ (b \circ c) = (a \circ b) \circ c$.
A3. (Identity) There exists an element i in S such that for all a in S,
$a \circ i = a$ and $i \circ a = a$.
A4. (Inverse) For each a in S there is an element denoted by a^{-1} (called a-inverse) such that $a \circ a^{-1} = i$ and $a^{-1} \circ a = i$.

There is a more general algebraic concept that satisfies only A1 and A2, the first half of the set of four axioms which has thus been called a *semigroup*.

When $(S, \circ)$ satisfies A1, A2, A3, it is known as a *semigroup with unit*, where the word "unit" is used as a synonym for identity element.

A different but equivalent axiom system for a group results when the last two axioms A3 and A4 are replaced by:

B3. (Left cancellation law)
If $x \circ a = x \circ b$, then $a = b$.

B4. (Right cancellation law)
If $a \circ y = b \circ y$, then $a = b$.

It is a standard exercise in group theory to prove first that every group satisfies statements B3 and B4, that is, that these are *theorems* about groups. It is then proved that in the system whose axioms are A1, A2, B3, B4, the

Appendix

statements A3 and A4 are theorems. This establishes the equivalence between the two axiom systems.

Another useful algebraic model is obtained from the axiom system A1, B3, B4. The result is called either a *quasigroup* or a *Latin square*.

One additional axiom is frequently added to A1–4 (or equivalently to A1, A2, B3, B4).

A5. (Commutativity)
     For all a, b in S, $a \circ b = b \circ a$.

If $(S, \circ)$ satisfies A1–5, it is called an *abelian group*. It is important to emphasize that not all groups are abelian, the smallest nonabelian group having six elements.

There are many other equivalent axiom systems for groups.

182

References

Abell, P. 1970. The structural balance of the kinship systems of some primitive peoples. In *Structuralism,* ed. M. Lane. New York, Basic Books.

Aberle, D. F. 1961. Navaho. In *Matrilineal kinship,* eds. D. M. Schneider and K. Gough. Berkeley, University of California Press.

Alkire, W. H. 1965. *Lamotrek atoll and inter-island socioeconomic ties.* Urbana, University of Illinois Press.

Alkire, W. H. 1977. *An introduction to the peoples and cultures of Micronesia.* Menlo Park, Calif., Cummings.

Allaire, F. 1980. Another proof of the Four Color Theorem. *Proceedings of the Seventh Manitoba Conference on Numerical Mathematics and Computing.* Winnipeg, University of Manitoba, no. 20, pp. 3–72.

Appel, K. and Haken, W. 1976. Every planar map is four colorable. *Bulletin of the American Mathematical Society,* vol. 82, pp. 711–12.

Armstrong, W. E. 1928. *Rossel Island.* Cambridge, Cambridge University Press.

Ascher, M. and Ascher, R. 1981. *Code of the Quipu.* Ann Arbor, University of Michigan Press.

Atkins, J. R. and Curtis, L. 1969. Game rules and the rules of culture. In *Game theory in the behavioral sciences,* eds. I. R. Buchler and H. G. Nutini. Pittsburgh, University of Pittsburgh Press.

Barnes, J. A. 1962. African models in the New Guinea Highlands. *Man,* vol. 62, pp. 5–9.

Barnes, J. A. 1969. Graph theory and social networks: A technical comment on connectedness and connectivity. *Sociology,* vol. 3, pp. 215–32.

Barnes, J. A. 1971. *Three styles in the study of kinship.* Berkeley, University of California Press.

Barnes, J. A. 1972. *Social networks.* Addison-Wesley Module in Anthropology.

Barnes, J. A. and Harary, F. 1983. Graph theory in network analysis. *Social Networks* (in press).

Barnes, R. H. 1982. Number and number use in Kédang, Indonesia. *Man* (n.s.), vol. 17, pp. 1–22.

Bateson, G. 1958. *Naven.* Stanford, Stanford University Press.

Beineke, L. W. and Harary, F. 1978a. Consistent graphs with signed points. *Rivista di Matematica per le Scienze Economiche e Sociale,* vol. I, pp. 81–8.

Beineke, L. W. and Harary, F. 1978b. Consistency in marked digraphs. *Journal of Mathematical Psychology,* vol. 18, pp. 260–9.

183

References

Berge, C. 1962. *The theory of graphs and its applications,* transl. A. Doig. New York, Wiley.

Berlin, B., Breedlove, D. E. and Raven, P. H. 1973. General principles of classification and nomenclature in folk biology. *American Anthropologist,* vol. 75, pp. 214–42.

Berlin, B. and Kay, P. 1969. *Basic color terms.* Berkeley, University of California Press.

Berndt, R. M. 1964. Warfare in the New Guinea Highlands. *American Anthropologist,* vol. 66, pp. 183–203.

Biggs, N. L., Lloyd, E. K. and Wilson, R. J. 1976. *Graph theory 1736–1936.* Oxford, Clarendon Press.

Bott, E. J. 1971. *Family and social network: Roles, norms and external relationships in ordinary urban families.* London, Tavistock.

Bower, G. H. 1970a. Analysis of a mnemonic device. *American Scientist,* vol. 58, pp. 496–510.

Bower, G. H. 1970b. Organizational factors in memory. *Cognitive Psychology,* vol. 1, pp. 18–46.

Brookfield, H. C. and Hart, D. 1971. *Melanesia: A geographical interpretation of an island world.* London, Methuen.

Brown, D. J. J. 1979. The structuring of Polopa feasting and warfare. *Man,* vol. 14, pp. 712–33.

Burton, M. L., Brudner, L. A. and White, D. R. 1977. A model of the sexual division of labor. *American Ethnologist,* vol. 4, pp. 227–51.

Carneiro, R. L. 1973. The four faces of evolution: Unilinear, universal, multilinear, and differential. In *Handbook of social and cultural anthropology,* ed. J. J. Honigmann. Chicago, Rand McNally.

Carroll, M. P. 1973. Applying Heider's theory of cognitive balance to Claude Lévi-Strauss. *Sociometry,* vol. 36, pp. 285–301.

Carroll, M. P. 1977. Leach, genesis and structural analysis: A critical evaluation. *American Ethnologist,* vol. 4, pp. 663–77.

Cartwright, D. and Gleason, T. C. 1967. A note on a matrix criterion for unique colorability of a signed graph. *Psychometrika,* vol. 32, pp. 291–6.

Cartwright, D. and Harary, F. 1956. Structural balance: A generalization of Heider's theory. *Psychological Review,* vol. 63, pp. 277–93.

Cartwright, D. and Harary, F. 1959. A note on Freud's "Instincts and their vicissitudes." *International Journal of Psychoanalysis,* vol. 40, pp. 287–90.

Cartwright, D. and Harary, F. 1968. On the coloring of signed graphs. *Elemente der Mathematik,* vol. 23, pp. 85–9.

Cartwright, D. and Harary, F. 1979. Balance and clusterability: An overview. In *Perspectives on social network research,* ed. S. Leinhardt. New York, Academic Press.

Cayley, A. 1891. On the analytical forms called trees, with application to the theory of chemical combinations. *Mathematical Papers,* vol. 10, Cambridge.

Cicero, M. T. 1942. *De oratore,* transl. E. W. Sutton and H. Rackham. Cambridge, Harvard University Press.

Clarke, D. L. 1968. *Analytical archaeology.* London, Methuen.

Cohen, J. E. 1978. *Food webs and niche space.* Princeton, N. J., Princeton University Press.

References

Conklin, H. C. 1964. Ethnogenealogical method. In *Explorations in cultural anthropology*, ed. W. Goodenough. New York, McGraw-Hill.

Crowe, D. W. 1975. The geometry of African art. II, A catalogue of Benin patterns. Historia Mathematica, vol. 2, pp. 253–71.

Crump, T. 1980. Trees and stars: Graph theory in southern Mexico. In *Numerical techniques in social anthropology*, ed. J. C. Mitchell. Philadelphia, Institute for the Study of Human Issues.

D'Andrade, R. G. 1976. A propositional analysis of U.S. American beliefs about illness. In *Meaning in anthropology*, eds. K. H. Basso and H. A. Selby. Albuquerque, University of New Mexico Press.

Davis, J. A. 1963. Structural balance, mechanical solidarity and interpersonal relations. *American Journal of Sociology*, vol. 68, pp. 444–62.

Davis, J. A. 1967. Clustering and structural balance in graphs. *Human Relations*, vol. 20, pp. 181–7.

Deacon, A. B. 1934. Geometrical drawings from Malekula and other islands of of the New Hebrides. *Journal of the Royal Anthropological Institute*, vol. 64, pp. 129–75.

Doreian, P. 1971. *Mathematics and the study of social relations*. New York, Schocken.

Doreian, P. 1974. On the connectivity of social networks. *Journal of Mathematical Sociology*, vol. 3, pp. 245–58.

Doreian, P. 1981. Polyhedral dynamics and conflict mobilization in social networks. *Social Networks*, vol. 3, pp. 107–16.

Durkheim, E. and Mauss, M. 1963. *Primitive classification*, transl. R. Needham. Chicago, University of Chicago Press.

Euler, L. 1736. Solutio problematis ad geometriam situs pertinentis. *Commentarii Academiae Scientiarum Imperialis Petropolitanae*, vol. 8, pp. 128–40.

Evans-Pritchard, E. E. 1929. The study of kinship in primitive societies. *Man*, vol. 29, pp. 190–4.

Feibleman, J. and Friend, J. W. 1945. The structure and function of organization. *Philosophical Review*, vol. 54, pp. 19–44.

Feller, W. 1950. *An introduction to probability theory and its applications*. Vol. 1. New York, Wiley.

Firth, R. 1957. *We the Tikopia*. London, George Allen and Unwin.

Flament, C. 1963. *Applications of graph theory to group structure*, transl. M. Pinard, R. Breton, and F. Fontaine. Englewood Cliffs, N.J., Prentice-Hall.

Ford, L. R. and Fulkerson, D. R. 1957. A simple algorithm for finding maximal network flows and an application to the Hitchcock problem. *Canadian Journal of Mathematics*, vol. 9, pp. 210–18.

Ford, L. R. and Fulkerson, D. R. 1962. *Flows in networks*. Princeton, N.J., Princeton University Press.

Forge, A. 1972. The golden fleece. *Man* (n.s), vol. 7, pp. 527–40.

Fortes, M. 1969. *Kinship and the social order*. Chicago, Aldine.

Fortune, R. F. 1942. *Arapesh*. New York, J. J. Augustine.

Frank, O. and Harary, F. 1980. Balance in stochastic signed graphs. *Social Networks*, vol. 2, pp. 155–63.

Freedman, M. P. 1970. Social organization of a Siassi island community. In

References

Cultures of the Pacific, eds. T. G. Harding and B. J. Wallace. New York, Free Press.

Freeman, L. C. 1977. A set of measures of centrality based on betweenness. *Sociometry,* vol. 40, pp. 35–41.

Freeman, L. C. 1979. Centrality in social networks I: Conceptual clarification. *Social Networks,* vol. 1, pp. 215–39.

Freud, S. 1913. *The interpretation of dreams,* transl. A. Brill. New York, Macmillan.

Freud, S. 1925. Instincts and their vicissitudes. *Collected Papers,* vol. 4. London, Hogarth.

Freud, S. 1950. *Totem and taboo,* transl. J. Strachey. New York, Norton.

Freud, S. 1959. On the sexual theories of children. *Collected Papers,* vol. 2, transl. J. Riviere. New York, Basic Books.

Freud, S. 1960. *The ego and the id,* transl. J. Strachey. New York, Norton.

Freud, S. 1961. *Beyond the pleasure principle,* transl. J. Strachey. New York, Norton.

Freud, S. 1964. *New introductory lectures,* transl. J. Strachey. New York, Norton.

Friedell, M. F. 1967. Organizations as semilattices. *American Sociological Review,* vol. 32, pp. 46–54.

Gandy, R. 1973. "Structure" in mathematics. In *Structuralism, an introduction,* ed. D. Robey. Oxford, Clarendon Press.

Gardner, M. 1974. The combinatorial basis of the "I Ching" the Chinese book of divination and wisdom. *Scientific American,* vol. 230 (1), pp. 108–13.

Gardner, M. 1980. The capture of the monster: A mathematical group with a ridiculous number of elements. *Scientific American,* vol. 242 (6), pp. 20–32.

Geertz, C. 1973. Person, time and conduct in Bali. In *The interpretation of cultures.* New York, Basic Books.

Gifford, E. W. 1929. *Tongan society.* Honolulu, B. P. Bishop Museum Bulletin 61.

Gladwin, T. 1970. *East is a big bird.* Cambridge, Harvard University Press.

Gluckman, M. 1960. *Order and rebellion in tribal Africa.* Glencoe, Ill., Free Press.

Goldman, I. 1970. *Ancient Polynesian society.* Chicago, University of Chicago Press.

Goodenough, W. H. 1969. Rethinking "status" and "role": Toward a general model of the cultural organization of social relationships. In *Cognitive anthropology,* ed. S. A. Tyler. New York, Holt, Rinehart and Winston.

Greenberg, J. H. 1963. Some universals of grammar with particular reference to the order of meaningful elements. In *Universals of language,* ed. J. H. Greenberg. Cambridge, MIT Press.

Greenberg, J. H. 1978. Typology and cross-linguistic generalizations. In *Universals of human language,* vol. 1, Method and theory, ed. J. H. Greenberg. Stanford, Calif., Stanford University Press.

Greimas, A. J. 1970. *Du sens.* Paris, Seuil.

Hage, P. 1973. A graph theoretic approach to the analysis of alliance structure and local grouping in Highland New Guinea. *Anthropological Forum,* vol. 3, pp. 280–94.

Hage, P. 1976a. Structural balance and clustering in Bushmen kinship relations. *Behavioral Science,* vol. 21, pp. 36–37.

References

Hage, P. 1976b. The atom of kinship as a directed graph. *Man* (n.s.) vol. 11, pp. 558–68.

Hage, P. 1978. Speculations on Puluwatese mnemonic structure. *Oceania,* vol. 49, pp. 81–95.

Hage, P. 1979a. Graph theory as a structural model in cultural anthropology. *Annual Review of Anthropology,* vol. 8, pp. 115–36.

Hage, P. 1979b. A further application of matrix analysis to communication structure in Oceanic anthropology. *Mathématique et Sciences Humaines,* vol. 17, pp. 51–69.

Hage, P. 1979c. A Viennese autochthonous hero: Structural duality in Freud's origin myths. *Social Science Information,* vol. 18, pp. 307–24.

Hage, P. 1979d. Symbolic culinary mediation: A group model. *Man* (n.s.), vol. 14, pp. 81–92.

Hage, P. 1981. On male initiation and dual organization in New Guinea. *Man* (n.s.), vol. 16, pp. 268–75.

Hage, P. and Harary, F. 1981a. Mediation and power in Melanesia. *Oceania,* vol. 52, pp. 124–35.

Hage, P. and Harary, F. 1981b. Pollution beliefs in Highland New Guinea. *Man* (n.s.), vol. 16, pp. 367–75.

Hage, P. and Harary, F. 1982. On reciprocity in kinship relations. *Cambridge Anthropology,* vol. 9, pp. 39–43.

Hage, P. and Harary, F. 1983. Arapesh sexual symbolism, primitive thought and boolean groups. *L'Homme* (in press).

Hallpike, C. R. 1970. The principles of alliance formation between Konso towns. *Man* (n.s.), vol. 5, pp. 258–80.

Hallpike, C. R. 1976. Is there a primitive mentality? *Man* (n.s.), vol. 11, pp. 253–270.

Hallpike, C. R. 1979. *The foundations of primitive thought.* Oxford, Clarendon Press.

Hammond, N. D. C. 1972a. Locational models and the site of Lubaatún: A classic Maya centre. In *Models in Archaeology,* ed. D. L. Clarke. London, Methuen.

Hammond, N. D. C. 1972b. The planning of a Maya ceremonial center. *Scientific American,* vol. 226, pp. 83–91.

Hampton, J. A. 1982. A demonstration of intransitivity in natural categories. *Cognition,* vol. 12, pp. 151–64.

Hanby, J. P. 1974. The social nexus: Problems and solutions in the portrayal of primate social structures. *Symposium of the Fifth Congress of the International Primatological Society.* Basel, S. Karger.

Harary, F. 1953. On the notion of balance of a signed graph. *Michigan Mathematical Journal,* vol. 2, pp. 143–6.

Harary, F. 1955. On local balance and N-balance in signed graphs. *Michigan Mathematical Journal,* vol. 3, pp. 37–41.

Harary, F. 1957. Structural duality. *Behavioral Science,* vol. 2, pp. 255–65.

Harary, F. 1959a. On the measurement of structural balance. *Behavioral Science,* vol. 4, pp. 316–23.

Harary, F. 1959b. Status and contrastatus. *Sociometry,* vol. 22, pp. 23–43.

Harary, F. 1961a. A structural analysis of the situation in the Middle East in 1956. *Journal of Conflict Resolution,* vol. 5, pp. 167–78.

References

Harary, F. 1961b. A very independent axiom system. *American Mathematical Monthly,* vol. 68, pp. 159–62.

Harary, F. 1961c. A parity relation partitions its field distinctly. *American Mathematical Monthly,* vol. 68, pp. 215–17.

Harary, F. 1961d. Who eats whom? *General Systems,* vol. 6, pp. 41–4.

Harary, F. 1966. Structural study of 'A Severed Head.' *Psychological Reports,* vol. 19, pp. 473–4.

Harary, F. 1969 *Graph theory.* Reading, Mass., Addison-Wesley.

Harary, F. 1983. Consistency theory is alive and well. *Personality and Social Psychology Bulletin,* vol. 9, pp. 60–4.

Harary, F. and Kommel, H. J. 1979. Matrix measures for transitivity and balance. *Journal of Mathematical Sociology,* vol. 6, pp. 199–210.

Harary, F. and Norman, R. Z. 1953. *Graph theory as a mathematical model in social science.* University of Michigan Institute for Social Research.

Harary, F., Norman, R. Z. and Cartwright, D. 1965. *Structural models: An introduction to the theory of directed graphs.* New York, Wiley.

Harary, F. and Ostrand, P. A. 1971. The cutting center theorem for trees. *Discrete Mathematics,* vol. 1, pp. 7–18.

Harary, F. and Palmer, E. M. 1973. *Graphical enumeration.* New York, Academic Press.

Harary, F., Palmer, E. M., Robinson, R. W. and Schwenk, A. J. 1977. Enumertion of graphs with signed points and lines. *Journal of Graph Theory,* vol. 1, pp. 295–308.

Harary, F. and Schwenk, A. J. 1973. The number of caterpillars. *Discrete Mathematics,* vol. 6, pp. 359–65.

Harary, F. and Schwenk, A. J. 1974a. Efficiency of dissemination of information in one-way and two-way communication networks. *Behavioral Science,* vol. 19, pp. 133–5.

Harary, F. and Schwenk, A. J. 1974b. The communication problem on graphs and digraphs. *Journal of the Franklin Institute,* vol. 297, pp. 491–5.

Healy, B. and Stein, A. 1973. The balance of power in international history: Theory and reality. *Journal of Conflict Resolution,* vol. 17, pp. 33–61.

Heider, F. 1946. Attitudes and cognitive organization. *Journal of Psychology,* vol. 21, pp. 107–12.

Heider, F. 1958. *The psychology of interpersonal relations.* New York, Wiley.

Henley, N. M., Horsfall, R. B. and De Soto, C. B. 1969. Goodness of figure and social structure. *Psychological Review,* vol. 76, pp. 194–204.

Henry, J. 1954. The formal social structure of a psychiatric hospital. *Psychiatry,* vol. 17, pp. 139–51.

Hymes, D. H. 1971. The "wife" who "goes out" like a man: Reinterpretation of a Clackamas Chinook myth. In *Essays in semiotics,* eds. J. Kristeva, J. Rey-Debove and D. J. Umiker. The Hague, Mouton.

The I Ching or Book of changes, transl. R. Wilhelm and C. F. Baynes, 1967. Princeton, N.J., Princeton University Press.

Inhelder, B. and Piaget, J. 1958. *The growth of logical thinking,* transl. A. Parsons and S. Milgram. New York, Basic Books.

References

Irwin, G. J. 1974. The emergence of a central place in coastal Papuan prehistory: A theoretical approach. *Mankind,* vol. 9, pp. 268–72.

Irwin, G. J. 1978. Pots and entrepôts: A study of settlement, trade and the development of economic specialization in Papuan prehistory. *World Archaeology,* vol. 9, pp. 299–319.

Jordan, C. 1869. Sur les assemblages de lignes. *Journal für die reine und angewandte Mathematik,* vol. 70, pp. 185–90.

Kainen, P. C. and Saaty, T. L. 1977. *The four-color problem, assaults and conquest.* New York, McGraw Hill.

Kapferer, B. 1969. Norms and the manipulation of relationships in a work context. In *Social networks in urban situations,* ed. J. C. Mitchell. Manchester, Manchester University Press.

Kay, P. 1966. Comment on "Ethnographic semantics: A preliminary survey" by B. N. Colby. *Current Anthropology,* vol. 7, pp. 20–3.

Kay P. 1975. A model-theoretic approach to folk taxonomy. *Social Science Information,* vol. 14, pp. 151–66.

Kelly, R. C. 1974. *Etoro social structure.* Ann Arbor, University of Michigan Press.

Kemeny, J. G. and Snell, J. L. 1960. *Finite Markov chains.* New York, Van Nostrand.

Kemeny, J. G., Snell, J. L. and Thompson, G. L. 1956. *Introduction to finite mathematics.* Englewood Cliffs, N. J., Prentice-Hall.

Kendall, M. G. and Smith, B. B. 1940. On the method of paired comparisons. *Biometrika,* vol. 31, pp. 324–45.

König, D. 1936. *Theorie der endlichen und unendlichen Graphen.* Leipzig, Akademische Verlagsgesellschaft M.B.H. (Reprinted Chelsea, New York, 1950.)

Kuper, A. 1979. A structural approach to dreams. *Man* (n.s.), vol. 14, pp. 645–62.

Labby, D. 1976. *The demystification of Yap.* Chicago, University of Chicago Press.

Landau, H. G. 1951. On dominance relations and the structure of animal societies: Effect of inherent characteristics. *Bulletin of Mathematical Biophysics,* vol. 13, pp. 1–19, 245–62.

Leach, E. R. 1961. *Rethinking anthropology.* London, Athlone Press.

Leach, E. R. 1970. *The legitimacy of Solomon,* ed. M. Lane. New York, Basic Books.

Lemaître, Y. 1970. Les relations inter-insulaires traditionelles en Océanie: Tonga. *Journal de la Société des Océanistes,* vol. 26, pp. 93–105.

Lessa, W. A. 1950. Ulithi and the outer native world. *American Anthropologist,* vol. 52, pp. 27–52.

Lessa, W. A. 1966. *Ulithi: A Micronesian design for living.* New York, Holt, Rinehart and Winston.

Levison, M., Ward, R. G. and Webb, J. W. 1973. *The settlement of Polynesia.* Minneapolis, University of Minnesota Press.

Lévi-Strauss, C. 1955. The mathematics of man. *International Social Science Bulletin,* vol. 6, pp. 581–90.

Lévi-Strauss, C. 1960. On manipulated sociological models. *Bijdragen tot de Taal-, Land- en Volkenkunde,* vol. 116, pp. 45–54.

Lévi-Strauss, C. 1963a. Social structure. In *Structuralism,* transl. C. Jacobson and B. G. Schoepf. New York, Basic Books.

References

Lévi-Strauss, C. 1963b. *Totemism,* transl. R. Needham. Boston, Beacon Press.

Lévi-Strauss, C. 1963c. Structural analysis in linguistics and in anthropology. In *Structural anthropology,* transl. C. Jacobson and B. G. Schoepf. New York, Basic Books.

Lévi-Strauss, C. 1963d. The structural study of myth. In *Structural anthropology,* transl. C. Jacobson and B. G. Schoepf. New York, Basic Books.

Lévi-Strauss, C. 1966. The culinary triangle. *New Society* (London) (22 December), vol. 166, pp. 937–40.

Lévi-Strauss, C. 1968. *The origin of table manners,* transl. J. and D. Weightman. New York, Harper & Row.

Lévi-Strauss, C. 1969. *The elementary structures of kinship.* Revised edition, transl. J. H. Bell and J. R. von Sturmer. Boston, Beacon Press.

Lévi-Strauss, C. 1973. *From honey to ashes,* transl. J. and D. Weightman. New York, Harper & Row.

Lévi-Strauss, C. 1975. *The raw and the cooked,* transl. J. and D. Weightman. New York, Harper & Row.

Lévi-Strauss, C. 1976a. Reflections on the atom of kinship. In *Structural anthropology,* vol. 2, transl. M. Layton. New York, Basic Books.

Lévi-Strauss, C. 1976b. Structure and form: Reflections on a work by Vladimir Propp. In *Structural anthropology,* vol. 2. New York, Basic Books.

Lévi-Strauss, C. 1981. *Naked man,* transl. J. and D. Weightman. New York, Harper & Row.

Lewis, D. 1978. *The voyaging stars.* Sydney, Fontana Collins.

Linde, C. and Labov, W. 1975. Spatial networks as a site for the study of language and thought. *Language,* vol. 51, pp. 924–39.

Lingenfelter, S. G. 1975. *Yap: Political leadership and culture change in an island society.* Honolulu, University of Hawaii Press.

Livingstone, F. B. 1969. The application of structural models to marriage systems in anthropology. In *Game theory in the behavioral sciences,* eds. I. R. Buchler and H. G. Nutini. Pittsburgh, University of Pittsburgh Press.

Lynch, K. 1960. *The image of the city.* Cambridge, MIT Press.

Malinowski, B. 1961. *Argonauts of the Western Pacific.* New York, Dutton.

Marriott, M. 1968. Caste ranking and food transactions: A matrix analysis. In *Structure and change in Indian society,* eds. M. Singer and B. S. Cohn. Chicago, Aldine.

Mauss, M. 1954. *The gift: Forms and functions of exchange in archaic societies,* transl. I. Cunnison. London, Cohen and West.

Maybury-Lewis, D. 1979. Cultural categories of the Central Gê. In *Dialectical societies,* ed. D. Maybury-Lewis. Cambridge, Harvard University Press.

Mead, M. 1938. *The Mountain Arapesh I. An importing culture.* American Museum of Natural History, Anthropological Papers, Vol. 36.

Mead, M. 1940. *The Mountain Arapesh II. Supernaturalism.* American Museum of Natural History, Anthropological Papers, vol. 37.

Mead, M. 1947. *The Mountain Arapesh III. Socioeconomic life; IV. Diary of events in Alitoa.* American Museum of Natural History, Anthropological Papers, vol. 40.

Mead, M. 1961. The Arapesh of New Guinea. In *Cooperation and competition,* ed. M. Mead. Boston, Beacon Press.

References

Mead, M. 1963. *Sex and temperament in three primitive societies.* New York, William Morrow.

Menger, K. 1927. Zur allgemeinen Kurventheorie. *Fundamenta Mathematicae,* vol. 10, pp. 96–115.

Menger, K. 1981. On the origin of the N-arc theorem. *Journal of Graph Theory,* vol. 5, pp. 341–50.

Mitchell, J. C. 1969. The concept and use of social networks. In *Social networks in urban situations,* ed. J. C. Mitchell. Manchester, Manchester University Press.

Mitchell, J. C. 1980. Introduction to *Numerical techniques in social anthropology,* ed. J. C. Mitchell. Philadelphia. Institute for the Study of Human Issues.

Moon, J. W. 1963. An extension of Landau's theorem on tournaments. *Pacific Journal of Mathematics,* vol. 13, pp. 1343–5.

Murdoch, I. 1961. *A severed head.* New York, Viking Press.

Murdock, G. P. 1967. *Ethnographic atlas.* Pittsburgh, University of Pittsburgh Press.

Needham, J. 1956. *Science and civilisation in China,* vol. 2. Cambridge, Cambridge University Press.

Needham, J. 1978. *The shorter science and civilisation in China,* vol. 1. An abridgement by C. A. Ronan. Cambridge, Cambridge University Press.

Needham, R. 1975. Polythetic classification: Convergence and consequences. *Man* (n.s.), vol. 10, pp. 349–69.

Needham, R. 1979. *Symbolic classification.* Santa Monica, Calif., Goodyear.

Neisser, U. 1976. *Cognition and reality.* San Francisco, W. H. Freeman.

Oliver, D. L. 1955. *A Solomon Island society.* Boston, Beacon Press.

Olnick, M. 1978. *An introduction to mathematical models in the social and life sciences.* Reading, Mass., Addison-Wesley.

Oppitz, M. 1975. *Notwendige Beziehungen.* Frankfurt am Main, Suhrkamp Verlag.

Ott, S. 1981. *The circle of mountains.* Oxford, Clarendon Press.

Peay, E. R. 1976. A note concerning the connectivity of social networks. *Journal of Mathematical Sociology,* vol. 4, pp. 319–21.

Peirce, C. S. 1931. *Collected papers,* eds. C. Hartshorne and P. Weiss. Cambridge, Harvard University Press.

Piaget, J. 1957. *Logic and psychology,* transl. W. Mays. New York, Basic Books.

Piaget, J. 1960. *The psychology of intelligence,* transl. M. Piercy and D. E. Berlyne. Totowa, N.J., Littlefield, Adams.

Piaget, J. 1971. *Structuralism,* transl. and ed. C. Maschler. New York, Harper & Row.

Pitts, F. R. 1965. A graph theoretic approach to historical geography. *Professional Geographer,* vol. 17 (5), pp. 15–20.

Pitts, F. R. 1979. The medieval river trade network of Russia revisited. *Social Networks,* vol. 1, pp. 285–92.

Pouwer, J. 1966. Towards a configurational approach to society and culture in New Guinea. *Journal of the Polynesian Society,* vol. 75, pp. 267–86.

Propp, V. 1968. *Morphology of the folktale,* transl. L. Scott. Austin, University of Texas Press.

Quintilianus, M. F. 1921. *The Institutio oratoria of Quintilian,* transl. H. E. Butler. New York, Putnam.

References

Randall, R. A. 1976. How tall is a taxonomic tree? Some evidence for dwarfism. *American Ethnologist*, vol. 3, pp. 543–53.

Rappaport, R. A. 1968. *Pigs for the ancestors*. New Haven, Yale University Press.

Read, K. E. 1951. The Gahuku-Gama of Central Highlands. *South Pacific*, vol. 5, pp. 154–64.

Read, K. E. 1954. Cultures of the Central Highlands, New Guinea. *Southwestern Journal of Anthropology*, vol. 10, pp. 1–43.

Regnier, A. 1971. De la théorie des groupes à la pensée sauvage. In *Anthropologie et calcul*, eds. P. Richard and R. Jaulin. Paris, Union Générale d'Editions.

Rhetorica ad Herennium, transl. H. Caplan 1954. Cambridge, Harvard University Press.

Riesenberg, S. H. 1972. The organisation of navigational knowledge on Puluwat. *Journal of the Polynesian Society*, vol. 81, pp. 19–56.

Roberts, F. S. 1976. *Discrete mathematical models*. Englewood Cliffs, N.J., Prentice-Hall.

Roberts, F. S. 1979. Structural modeling and measurement theory. *Technological Forecasting and Social Change*, vol. 14, pp. 353–65.

Roheim, G. 1932. The psychoanalysis of cultural types. *International Journal of Psychoanalysis*, vol. 13, pp. 1–224.

Russell, B. 1901. Recent work on the principles of mathematics. *International Monthly*, vol. 4 (July), pp. 83–101.

Ryan, D. 1959. Clan formation in the Mendi valley. *Oceania*, vol. 29, pp. 257–89.

Ryder, J. W. and Blackman, M. B. 1970. The avunculate: A cross-cultural critique of Claude Lévi-Strauss. *Behavioral Science Notes*, vol. 5, pp. 97–115.

Sade, D. A. F. de. 1967. Les cent vingt journées de sodome. In *Oeuvres complètes*. Paris, Jean-Jacques Pauvert.

Sahlins, M. D. 1963. Poor man, rich man, big man, chief: political types in Melanesia and Polynesia. *Comparative Studies in Society and History*, vol. 5, pp. 285–303.

Sahlins, M. D. 1968. *Tribesmen*. Englewood Cliffs, N.J., Prentice-Hall.

Schwimmer, E. 1973. *Exchange in the social structure of the Orokaiva*. New York, St. Martins.

Schwimmer, E. 1974. Objects of mediation: Myth and praxis. In *The unconscious in culture*, ed. I. Rossi. New York, Dutton.

Seligman, C. G. 1910. *The Melanesians of British New Guinea*. Cambridge, Cambridge University Press.

Service, E. R. 1962. *Primitive social organization*. New York, Random House.

Shannon, C. E. 1938. A symbolic analysis of relay and switching circuits. *Electrical Engineering (Transactions Supplement)*, vol. 57, pp. 713–23.

Shimbel, A. 1953. Structural parameters of communication networks. *Bulletin of Mathematical Biophysics*, vol. 15, pp. 501–57.

Shweder, R. A. 1982. On savages and other children. Review of C. R. Hallpike's *The foundations of primitive thought*. *American Anthropologist*, vol. 84, pp. 354–66.

Silberbauer, G. B. 1961. Aspects of the kinship system of the G/wi Bushmen of the Central Kalahari. *South African Journal of Science*, vol. 57, pp. 353–9.

References

Silberbauer, G. B. 1972. The G/wi Bushmen. In *Hunters and gatherers today,* ed. M. Bicchieri. New York, Holt, Rinehart and Winston.

Steward, J. H. 1938. *Basin-Plateau aboriginal sociopolitical groups.* Washington, D.C., Bureau of American Ethnology Bulletin No. 120.

Steward, J. H. 1955. *Theory of culture change.* Urbana, University of Illinois Press.

Steward, J. H. 1970. The foundations of Basin-Plateau Shoshonean society. In *Languages and cultures of Western North America,* ed. E. H. Swanson. Pocatello, Idaho State University Press.

Strathern, A. 1969. Descent and alliance in the New Guinea Highlands: Some problems of comparison. *Proceedings of the Royal Anthropological Institute for 1968,* pp. 37–52.

Strathern, A. 1971. *The rope of moka.* Cambridge, Cambridge University Press.

Sung, Z. D. 1934. *The symbols of Yi King.* Shanghai, The China Modern Education Company.

Sylvester, J. J. 1882. On the geometrical forms called trees. *Johns Hopkins University Circle,* vol. 1, 1879–82, pp. 202–3.

Taafe, E. J. and Gautier Jr., H. L. 1973. *Geography of transportation.* Englewood Cliffs, N.J., Prentice-Hall.

Taylor, H. F. 1970. *Balance in small groups.* New York, Van Nostrand Reinhold.

Thomas, D. H. 1972. A computer simulation model of Great Basin Shoshonean subsistence and settlement patterns. In *Models in archaeology,* ed. D. L. Clarke. London, Methuen.

Tversky, A. 1969. Intransitivity of preferences. *Psychological Review,* vol. 76, pp. 31–48.

Tyler, S. A. 1969. Introduction. In *Cognitive anthropology,* ed. S. A. Tyler. New York, Holt, Rinehart and Winston.

Ulam, S. 1951. On the Monte Carlo method. *Proceedings of the Second Symposium on Large Scale Digital Calculating Machinery.* Cambridge, Harvard University Press, pp. 207–12.

Veblen, O. 1922. *Analysis situs.* New York, American Mathematical Society.

Wallace, A. F. C. 1962. Culture and cognition. *Science,* vol. 135, pp. 351–7.

Washburn, D. K. 1977. A symmetry classification of Pueblo ceramic designs. In *Discovering past behavior: Experiments in the archaeology of the American Southwest,* ed. P. Grebinger. New York, Gordon and Breach.

Weil, A. 1969. On the algebraic study of certain types of marriage laws. Appendix to Part 1 of C. Lévi-Strauss, *The elementary structures of kinship.* Boston, Beacon Press.

Werbner, R. P. 1973. The superabundance of understanding: Kalanga rhetoric and domestic divination. *American Anthropologist,* vol. 75, pp. 1414–40.

Whitten, N. E. 1976. *Sacha Runa.* Urbana, University of Illinois Press.

Wilder, R. L. 1952. *Introduction to the foundations of mathematics.* New York, Wiley.

Williamson, R. W. 1937. *Religion and social organization in central Polynesia,* ed. R. Piddington. Cambridge, Cambridge University Press.

References

Wilson, P. J. 1980. *Man, the promising primate.* New Haven, Yale University Press.

Yates, F. A. 1966. *The art of memory.* Chicago, University of Chicago Press.

Young, M. W. 1971. *Fighting with food.* Cambridge, Cambridge University Press.

Zachary, W. W. 1977. An information flow model for conflict and fission in small groups. *Journal of Anthropological Research,* vol. 33, pp. 452–73.

Zajonc, R. B. and Burnstein, E. 1965. The learning of balanced and unbalanced social structures. *Journal of Personality,* vol. 33, pp. 153–63.

Index

195

Index

Index

versus structure, 38–9, 69
detour, 83
digraphs, 4, 68
  acyclic, 82
  strongly connected, 69
  unilaterally connected, 69
  weakly connected, 69
distance
  in a digraph, 69
  in a graph, 18
distance matrix, 109–10
  and betweenness centrality, 111–12
  and eccentricity of a point, 111
  and status and contrastatus, 111n
divination systems, 166–70
Doreian, P., 85, 88, 132, 135–9
duality principle, properties of, 116
  in set theory and propositional logic,
    116–17
Durkheim, E., 55, 166, 167

eccentricity, 30
ecology
  and digraphs of food chains, 91
  and markov chain model of forag-
    ing, 145–9
endpoint, 19
equivalence relation, 75, 178, 179, 180
Etoro, 65–6, 88, 97
Euclid's parallel postulate, 171–2
Euler, L., 10–12
Evans-Pritchard, E. E., 45, 56
exchange systems
  and blocks, 28–9
  centrality in, 34, 36–8
  and digraphs, 5
  and graphs, 3
  and networks, 7
  and signed graphs, 4

fahu, 81
Feibleman, J., 65, 76
Firth, R., 54–5
Five Element Theory, 89–91
Flament, C., 48, 49–50, 53, 111
fofofo/nibai, 4
folk mathematics, 9–10, 12
Ford, L. R., 132, 141, 144
forest, 22
Forge, A., 34
Fortes, M., 65
Four Color Conjecture, 42, 175

Frank, O., 64
Freeman, L. C., 31, 34–5, 38
Freud, S., 122, 123, 125–9
Friedell, M. F., 85
Friend, J. W., 65, 76
Fulkerson, D. R., 132, 141, 144

Gahuku-Gama, 56–60
Gandy, R., 153–4, 155
Gardner, M., 153, 167, 169
Gautier, H. L., 104
Geertz, C., 66n
Genesis as myth, 130
genetic epistemology, 151, 162–5
geodesic, 18
geodetic subgraph, 111–12
geometric art, of New Hebrides, 10, 12
Gifford, E. W., 80–1
Gilbert Islands, 164
Gladwin, T., 14, 89
Goldman, I., 80, 81
Goodenough, W. H., 81–2
graphical enumeration, 25
graphs, 3, 16
  complete, 18
  connected, 18
  eulerian, 12
  hamiltonian, 26
  labeled, 17
  n-chromatic, 41
  n-colorable, 41
  oriented, 87
  planar, 42
  regular, 17
Great Basin anthropology, 145–9
greatest lower bound, 85
Greenberg, J. H., 67
Greimas, A. J., 165–6
group models
  of culinary symbolism, 160–1
  in explanation of anomalous
    practices, 161–2
  in kinship attitudes, 48–9
  in myth, 7–8, 151–2
  and primitive thought, 9, 162–5
  and structuralist method, 8
groups, 154
  abelian, 182
  boolean, 155, 156, 161, 162
  cyclic, 155, 176
  isomorphic, 176
  Klein, 8, 48–9, 152, 155, 157, 161,

197

Index

Sylvester, J. J., 21, 30
symmetric relation, 71, 74, 178

Taafe, E. J., 104
teknonymy, 66n
theorems
    form of, 172
    use of, 9–12
Thomas, D. H., 132, 145–9
Tikopia, 54–5
Tonga, 80–3
tournaments, 79
    in kinship relations, 82
    transitive, 79
"tours," 25
trail, 17
transitive closure, 77–8
transitive relations, 71, 178
    in ethnosemantics, 73
    in kinship, 76, 82
    in subordination structures, 71–2
    vacuous, 75
transitivity ratio of a digraph, 106
transmitter, 83
trees, 15, 21
    caterpillars, 15, 22
    coding of, 23–4
    hydrocarbon, 25
    as mnemonic structures, 14–15,
        19–24
    oriented, 87
    as political structures, 88–9
triple
    cyclic, 80
    transitive, 80
Truk, 80–3

Tversky, A., 73
Tyler, S. A., 24

Ulam, S., 147
unary operation, 120
universal matrix, 105
universal relation, 177, 178, 179, 180
upper bound, 85

value matrix, 133–4

walk
    closed, in a digraph, 68
    closed, in a graph, 17
    in a digraph, 68
    in a graph, 17
    open, in a digraph, 68
    open, in a graph, 17
    spanning, 69
Wallace, A. F. C., 71
Ward, R. G., 94, 95
Webb, J. W., 94, 95
weight at a point, 30
Weil, A., 25
Werbner, R. P., 166
White, D. R., 67
Whitten, N. E., 73–4
Wilson, P. J., 67

Yap, 60, 61, 88–9
Yates, F. A., 19
Young, M. W., 4

Zachary, W. W., 132, 141–4
Zajonc, R. B., 20
Zambia, 132, 135, 138

CAMBRIDGE STUDIES IN SOCIAL ANTHROPOLOGY

Editor: Jack Goody

*Available as a paperback

CPSIA information can be obtained at www.ICGtesting.com
Printed in the USA
LVOW08s1743200516

489272LV00001B/120/P